Wireless Markup Language (WML)

Scripting and Programming using WML, cHTML, and xHTML

By

Bill Routt

Published By:
Althos Publishing
404 Wake Chapel Road
Fuquay-Varina, NC 27526 USA
Telephone: 1-800-227-9681
Fax: 1-919-557-2261
email: Success@Althos.com
web: www.Althos.com

Althos

All rights reserved. No part of this book may be reproduced or transmitted in any form or by any means, electronic or mechanical, including photocopying recording or by any information storage and retrieval system without written permission from the authors and publisher, except for the inclusion of brief quotations in a review.

Copyright © 2004 By Althos Publishing
First Printing

Printed and Bound by Lightning Source

> Every effort has been made to make this manual as complete and as accurate as possible. However, there may be mistakes both typographical and in content. Therefore, this text should be used only as a general guide and not as the ultimate source of information. Furthermore, this manual contains information on telecommunications accurate only up to the printing date. The purpose of this manual to educate. The authors and Althos Publishing shall have neither liability nor responsibility to any person or entity with respect to any loss or damage caused, or alleged to be caused, directly or indirectly by the information contained in this book.

International Standard Book Number: 0-9742787-5-0

Acknowledgements

We thank the many gifted people who gave their technical and emotional support for the creation of this book. In many cases, published sources were not available on this subject area. Experts from manufacturers, service providers, trade associations and other telecommunications related companies gave their personal precious time to help us and for this we sincerely thank and respect them.

We thank the industry experts including:Sergio Rogina with Samsung, Andrew Tingley from Research in Motion (RIM), Brenda Ning at Sony Ericsson, Nita Miller with HP Invent, and Keith Nowak with Nokia and Ben Forta from Macromedia, Inc., Pekka Niskanen of Sybex Verlag GmbH, and H.M. Deitel of Deitel and Associates for providing material that helped the development of this book:

Special thanks to the people who assisted with the production of this book including: Lawrence Harte (Director), Tom Pazderka (illustator), Karen Bunn (Project Manager), and Katie Jackson (editor).

About the Author

Mr. Routt is a communications product and technology expert and is the department head of Electronics Engineering Technology and Automation /Robotics technology at Wake Technology Community College. Mr. Routt has over 33 years of technical, research, design, development, and instruction experience. He has worked for leading companies including Bell Laboratories, DuPont, Modcomp, Siemens, Gould Computer, and several universities and colleges. Mr. Routt has been published in the Bell System Technical Journal and in the Bell Laboratories Record covering switching system technologies and telecommunications. He continually researches and develops software programs and creates courses on communications and technology automation. Mr. Routt holds many degrees and certificates including a MSEE from Carnegie Institute of Technology and a BSEE from the Pensylvania State University.

Wireless Markup Language (WML)

Dedications

"I wish to thank first and foremost my lovely wife, Trish, for initially giving me the idea, inspiration and confidence to write this book, and secondly for putting up with me being on the computer all the time during the writing of the book."

"Next, I wish to thank my oldest son, William J. Routt, for initially getting me interested in WML and the entire subject of programming for wireless devices. Bill is currently Director of Wireless Platform Testing for Sprint PCS, based in Kansas City, Mo."

"I also wish to thank Ben White, currently Vice-President of Curriculum Studies at Wake Technical Community College for allowing me to teach a course on Programming Wireless Devices within the Electronics Engineering Technology curriculum as an elective, multiple times."

Bill Routt
Author

Table of Contents

CHAPTER 1 - WIRELESS SYSTEMS AND APPLICATIONS   1
Cellular and PCS *1*
    Wireless Data *4*
    Paging *6*
    Wireless Languages *8*
    Wireless Access Protocol (WAP) *11*
    Wireless Markup Language (WML) *14*
    WMLScript *15*
    XHTML Basic *17*
  cHTML 19
  SUMMARY 20
  QUESTIONS 23

CHAPTER 2 - WIRELESS DEVICES AND OPERATING SYSTEMS   25
  MOBILE DEVICES 25
    Mobile Phones *26*
    Mobile Data Terminals *27*
    Two-Way Pagers *29*
    Personal Digital Assistants (PDAs) *29*
    Convergence Devices *30*
    Pocket PC *32*
  MOBILE OPERATING SYSTEMS AND PLATFORMS 34
    BREW *34*
    Palm Operating System (Palm OS) *35*
    EPOC *35*
    Windows CE *37*
    Stinger OS *37*

ix

Wireless Markup Language (WML)

MICROBROWSERS 38
SUMMARY 40
QUESTIONS 42

CHAPTER 3 - WIRELESS MARKUP LANGUAGE (WML) BASICS 43

BASIC STRUCTURE 43
INSERTING COMMENTS 46
VARIABLES 46
   Type Conversion of Variables *50*
TEXT FORMATTING 51
BASIC COMMANDS 56
   Card *56*
   Do *59*
   Go *61*
INPUT 64
   Anchor *68*
   Table *72*
   Prev *75*
   Noop *77*
SUMMARY 77
QUESTIONS 79
PROBLEMS 80

CHAPTER 4 - WML OBJECTS AND SYNTAX 83

TEMPLATES 83
PASSING VARIABLES 88
MORE COMMANDS 92
   Select *92*
   Option *95*
   Optgroup *96*
   Refresh *98*
   Timer *100*

Table of Contents

    Fieldset *101*
    Head *103*
    Meta *103*
    Access *104*
    Postfield *105*
    Onevent *106*
  Summary 107
  Questions 109
  Problems 110

CHAPTER 5 - WMLSCRIPT PROGRAMMING 111
  Scripting Structure 111
  Functions 112
  WMLScript Variables 115
  Operators 122
  Statements and Expressions 126
  Calling WMLScript Functions 137
  Summary 139
  Questions 141
  Problems 142

CHAPTER 6 - ADVANCED WML SCRIPT 143
  Type Conversions 143
  Standard Libraries 146
    String *148*
    Lang *154*
    Float *158*
    URL *161*
    WML Browser *164*
    Dialogs *167*

Wireless Markup Language (WML)

    Debug *170*
    Console *170*
  SUMMARY 171
  QUESTIONS 173
  PROBLEMS 174

CHAPTER 7 - SETTING UP A WAP SERVER 175
  NOKIA WAP SERVER 176
  MODIFYING AN HTTP SERVER 177
  PUBLISHING TO THE WAP SERVER 178
  SUMMARY 179
  QUESTIONS 180

CHAPTER 8 - CREATING PUSH AND PULL NOTIFICATIONS 181
  WHAT ARE NOTIFICATIONS 181
    Alerts *182*
    Cache Operations *183*
    Content Messages *183*
    Multi-part Messages *183*
  PUSH VERSUS PULL NOTIFICATIONS 184
  SUMMARY 187
  QUESTIONS 189

CHAPTER 9 - ADDING SECURITY TO APPLICATIONS .. 191
  SECURITY BASICS 191
    Threat Models *194*
  WAP SECURITY ARCHITECTURE 196
    Request Path *196*
    WTLS and SSL *199*
    Security Certificates *200*

Table of Contents

SESSION MANAGEMENT 201
   Client Authentication *202*
   WML for Secure Applications *203*
   Cleaning Up *204*
SUMMARY 205
QUESTIONS 208

CHAPTER 10 - OTHER SCRIPT LANGUAGES 209

XHTML 209
   Headers *212*
   Linking *214*
   Images *215*
   Special Characters *218*
   Tables *220*
   Unordered Lists *222*
   Nested and Ordered Lists *224*
   Simple XHTML Basic Forms *226*
   More Complex XHTML Basic Forms *229*
cHTML 233
   Headers *233*
   Images *236*
   Changing Text Color *238*
   Special Characters and More Line Breaks *239*
   Unordered Lists *241*
   Nested and Ordered Lists *242*
   cHTML Forms *244*
   Summary *245*
   Questions *248*
   Problems *250*

Wireless Markup Language (WML)

APPENDIX I 253

APPENDIX II 257

APPENDIX III 291

INDEX 267

Chapter 1

Wireless Systems and Applications

Wireless technology allows individuals and organizations to connect to the Internet and World Wide Web at any time, from almost any location, via wireless devices. Today's most popular wireless devices include cell phones, personal digital assistants (PDAs), pagers and laptops.

Wireless devices allow businesses and consumers to enhance their productivity in many different ways.

Cellular and PCS

A cellular network is composed of individual compartments, or cells, which combine to form the communications network. At the center of each cell is a base station that transmits analog or digital signals to and from users within that cell. A base station cannot communicate with users beyond the boundaries of its cell. If a mobile user passes through a cell boundary, the signal must be passed to the base station of the cell the user has entered. The transfer of signals from one base station to another is called a handoff. The handoff allows mobile users to continue communications without disconnections or network interruptions.

Cellular technology can be considered by first, second, or third generation technology. (1G, 2G, 3G) First-generation (1G) technology is the first form of wireless communications networks based on analog signals.

Wireless Markup Language (WML)

Second-generation (2G) technology includes most of the present technologies and will support many communications systems. (i.e., 2.5G, 3G, and 4G) 2G systems are designed to use digital-signal transmissions versus 1G analog-signal transmissions. 2G technologies include Personal Communications Systems (PCS), Global System for Mobile Communications (GSM), Time Division Multiple Access (TDMA), and Orthogonal Frequency Division Multiplexing (OFDM).

PCS was introduced in 1995 and uses the digital frequency spectrum. PCS can have both a general meaning encompassing digital wireless communications services, or it can also refer to the specific network that operates on the 1900 MHz frequency band used in wireless communication. Sprint PCS, one of the largest digital wireless service providers in the U.S., uses PCS technology. In general, PCS was designed to address the problems in cellular-based networks caused by strains on network capacity limits.

The structure of PCS networks is similar to analog networks, with cells and base stations used to complete communications. PCS breaks down the larger cell areas used by analog-based networks to form microcells. Microcells, also called picocells, allow more cells to inhabit one geographical area, thus providing better network coverage by reducing interference and increasing the chance that phone calls will have a clear reception and be completed without interruption.

Figure 1.1 shows the basic parts of a cellular mobile communication system. The mobile telephone has the ability to tune in to many different radio channel frequencies or codes. The base station commands the mobile telephone on which frequency to use in order to communicate with another base station that may be from two to fifteen miles away. The base station routes the radio signal to the MSC either by wire (e.g. a leased telephone line), microwave radio link, or fiberoptic line. The MSC connects the call to the public switched telephone network (PSTN) and the PSTN then connects the call to its designation (e.g. office telephone.)

Chapter 1

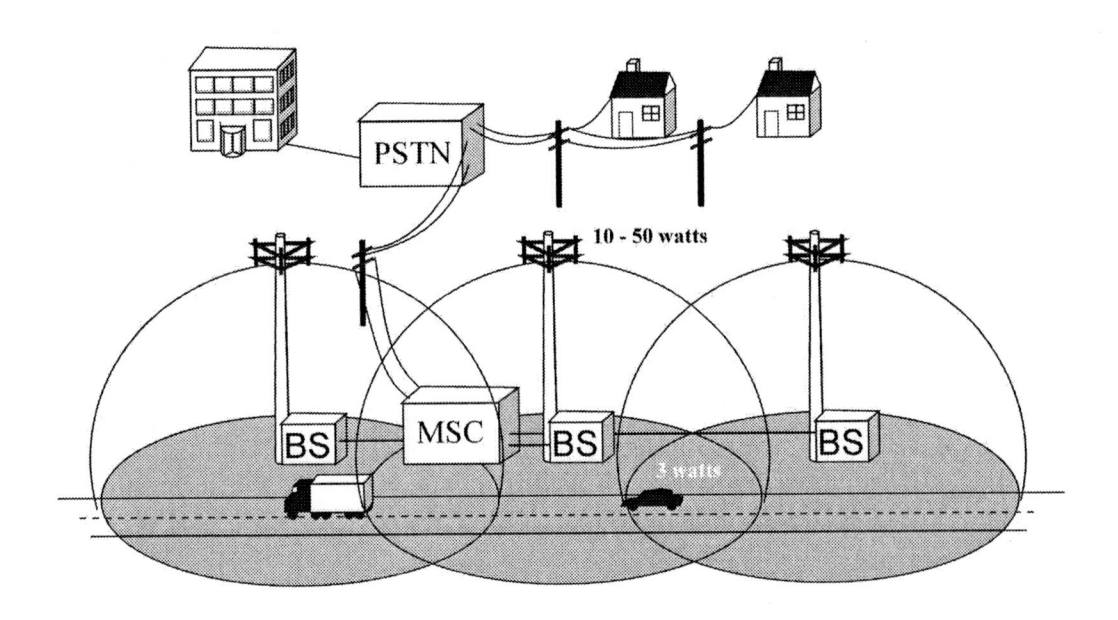

Figure 1.1 Cellular Network and PCS System

The Global System for Mobile Communications (GSM) is a PCS digital cellular communications network that accounts for 69.5 % of the digital wireless communications worldwide. GSM is used in 171 countries and offers data speeds of 9.6-14.4Kbps. GSM is a roaming technology, and allows users to make calls outside their home countries, but calls are usually expensive and coverage is often unreliable.

TDMA is the foundation for many 2G technologies and will be used in some 2.5G and 3G systems in the future. Many of the technologies in use today are part of the TDMA technology family because they are based on TDMA as an access technology or are used in combination with other access technologies like CDMA and FDMA. D-AMPS, also known as TDMA IS-136, is one of the most widely used networks today, especially in North America. Other regions of the world that have TDMA-based networks include Latin America, and the Middle East. As the technology moves to 2.5G and 3G, many varieties of TDMA systems could emerge that combine TDMA with other technologies like CDMA. But, some feel that TDMA could be phased out in favor of other technologies.

Orthogonal Frequency Division Multiplexing (OFDM) was first used for military and radio communications in the mid-1960s. OFDM is also used for DSL connections, radio, and TV broadcasts. AT&T Wireless is developing OFDM technology in its wireless networks. OFDM is used in the development of new technologies for wireless LANs and other networks as they move toward 3G, and possibly even 4G, systems. OFDM is based on Frequency Division Multiplexing (FDM). FDM divides frequency channels on the spectrum into smaller channels, or bands, which can be used multiple times. OFDM improves on the limitations of FDM by changing the signal shape and compacting more signals into one area of bandwidth to reduce the amount of spectrum in use.

Wireless Data

M-business, or Mobile-business, refers to e-business enabled by wireless communications and data. M-business is one of the newest frontiers in electronic communications. Though in its initial stages, m-business is growing rapidly. M-business has the ability to reach users effectively and allow them instant access to business-critical information and communications.

M-business differs significantly from e-business applications. Most wireless interfaces display only condensed text and basic graphics. Wireline communications, on the other hand, offer full-color, interactive, information-packed sites. Wireless transmission is also less secure than wireline transmissions.

Wireless access benefits businesses, employers, employees, and consumers. For employers and employees, wireless access provides the ability to communicate, access corporate databases, manage administrative tasks (such as answering e-mail and scheduling meetings), and enhance customer relations. Also, users can complete tasks during idle time- while waiting for a train, or standing in line at a bank.

Currently, accessing the Web through a wireless device is different than accessing the Web through wireline connections. Cell phone providers often charge by the minute, which is not economically conducive to surfing the Web via a wireless- enabled cell phone. Phone reception may be limited or unavailable, depending on the user's location when the call is made. Also, small, monochromatic cell phone screens typically have low resolution, and cell phone keypads cause difficulty in entering long strings of information. To reduce the inconvenience, some personal digital assistants (PDAs) use a stylus, a miniature keyboard, and handwriting recognition technology to simplify data entry. Other wireless devices and services further accommodate users by offering radio buttons, checkboxes, and drop-down lists for entering information.

In the future, 3G technologies will help foster Web surfing via wireless devices. These technologies will enable faster connection speeds and the ability to download streaming audio and video.

Figure 1.2 shows several types of wireless data systems and applications that use them. In this example, many types of wireless data devices communicate through a public and private wireless data systems. In the core of the system, there is a data packet switching system. The switching system commonly routes the data between the wireless device and a computer system (such as a company computer). Some wireless data systems use radio tower (base station) transmitters that can provide more than 500 watts effective radiated power (ERP) while portable mobile data devices can usually provide less than 1 watt of transmitted power. As a result, some wireless data systems require many more radio receiver sites than transmitter sites. This is required to allow low-power mobile data transmitters to reach the system.

Wireless Markup Language (WML)

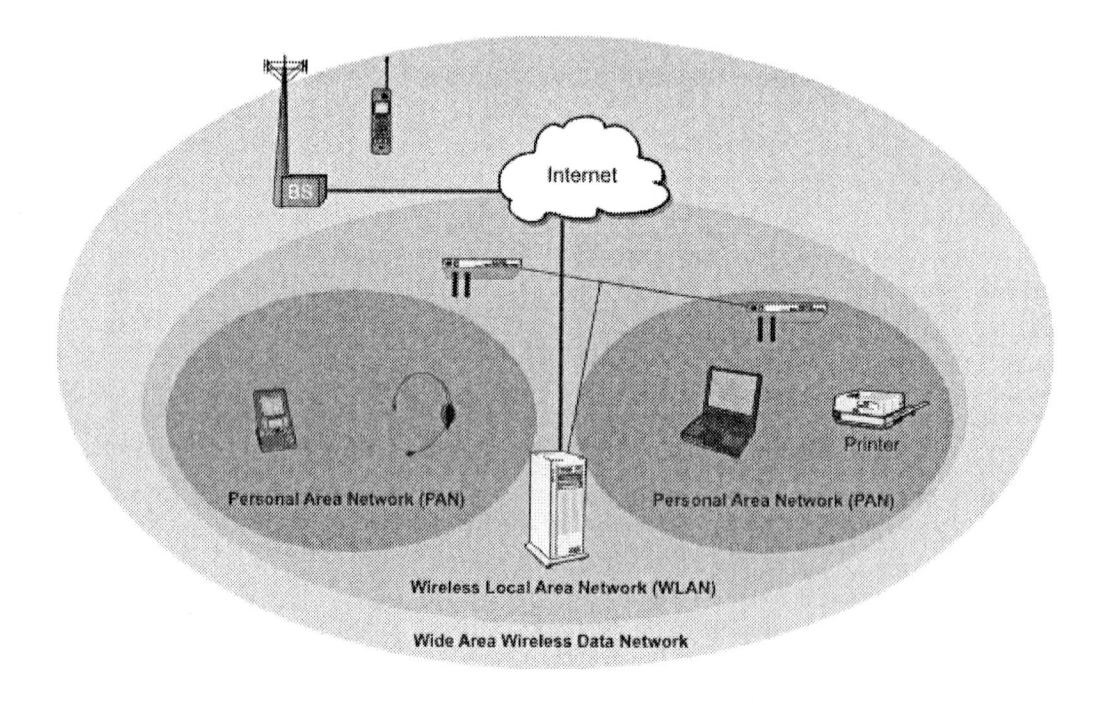

Figure 1.2 Wireless Data

Paging

Paging is accomplished via pagers, which were one of the first widely used devices for wireless communications. Pagers are designed to alert users when new messages arrive and show caller identification information and store messages. Similar to mobile phones, pagers vary in size, shape, color and functionality. There are a variety of paging service providers and pricing plans. Unlike mobile phone calling plans, pager plans do not charge on a per-minute basis. Pager plan rates are based on the number of messages sent and received and the number of pages stored per month.

Chapter 1

Pagers use three types of messaging technology. One-way numeric paging allows the transfer of numbers. People can use strings of numbers as codes to represent messages. Text and voice messages are not supported in this system.

Text-based messaging allows pagers to receive text and voice-mail messages. This one-way communication form is known as one-way text messaging. The most recent development in paging technology is two-way interactive communications. This technology allows the users to send and receive text messages by using handheld paging devices or mobile phones. Motorola, Timex, Philips, and Matsushita/Panasonic are leading manufacturers of paging and messaging products.

Mobile e-mail products and pages also are growing in popularity both for personal and professional use. Some companies provide e-mail solutions in addition to two-way interactive messaging (such as the BlackBerry handheld wireless device).

Figure 1.3 shows a typical two-way paging system. The two-way wireless pager transmits with typically 1 watt of power to one of the high gain receiving towers. The information is then sent over wires to the paging center and then either to the PSTN (Public Switched Telephone Network), to a satellite data link for a wide area paging system, or to the transmitting tower for local paging. In any case, when the return data (response) comes back to the Paging Center, it is sent to the transmitting tower that typically is 250-500 watt of power, and then to the two-way pager.

Wireless Markup Language (WML)

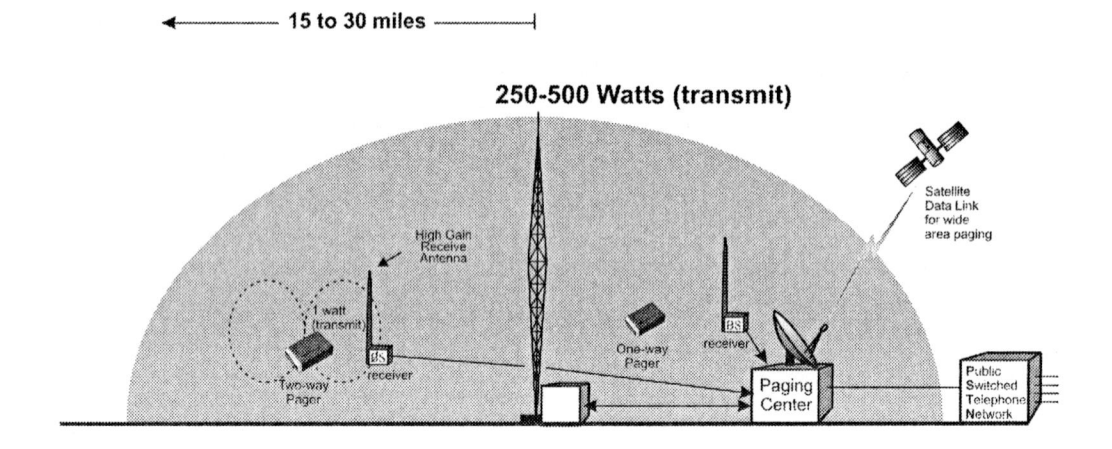

Figure 1.3 Two-Way Paging System

Wireless Languages

There are currently three main languages being used to program wireless devices, and each one is supported by different companies and consortiums.

1.) WML and WMLScript ,
2.) cHTML , and
3.) XHTML Basic.

WML is being used to program phones, pagers and PDA's. It was defined and developed by WAP (Wireless Application Protocol) Forum, which recently (June 2002) joined with the Open Mobile Architecture to form the OMA – Open Mobile Alliance. WML 1.1 was developed in 1999.

Chapter 1

cHTML was developed by NTT DoCoMo (a subsidiary of the Japanese firm NTT) in 1999, and is used to program the wireless devices that use i-mode, the wireless Internet access service provided by NTT DoCoMo.

XHTML Basic was developed as a language to program wireless devices by World Wide Web Consortium (W3C). It was initiated to provide a common markup language to use for wireless devices in December 2000.

Markup Family Tree:

The chart below shows how all of the markup languages are related, in terms of which ones were based on previous markup languages. As shown, HTML (Hypertext Markup Language) and HDML (Handheld Device Markup Language) were the original two markup languages. HTML was developed for use with PC-based web browsers, and as a language it is concerned with layout and placement of images on pages. HDML was developed for use with wireless handheld devices, and was developed by a company called Unwired Planet. In 1997 when the WAP Forum was formed, it created WML, replacing HDML.

Table 1.1 shows how the Extensible Markup Language (XML) is also as a root language to some of the currently used wireless markup languages. XML differs from WML or XHTML in that XML is a language for creating markup languages. XML allows document authors to create their own markup for virtually any type of information. Therefore, authors can use XML to create entirely new markup languages to describe specific types of data, including mathematical formulas, chemical formulas, music and recipes.

Wireless Markup Language (WML)

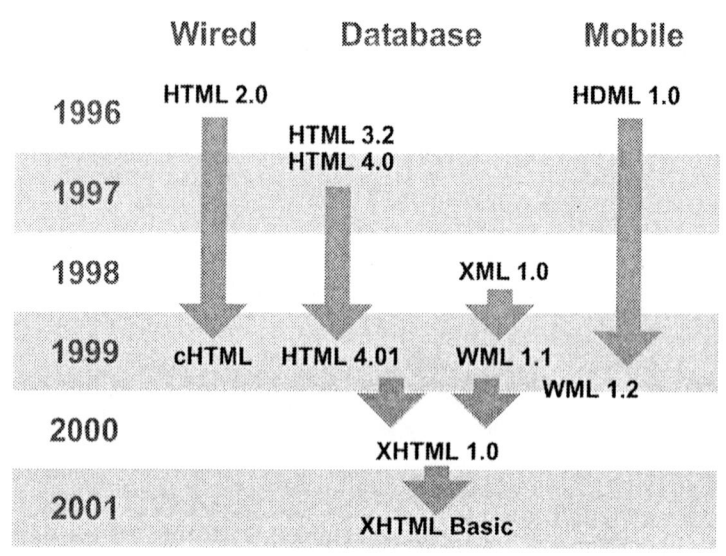

Table 1.1 Chart of the Markup Tree

Convergence?

While all three languages currently exist in the wireless world, OMA is trying to consolidate these languages and is developing a specification for WAP 2.0, which encompasses two of the three languages – XHTML Basic and WML 1. WAP 2.0 specifies WML 2, which encompasses XHTML Basic, and will be backward compatible to WML1. It utilizes standards developed by W3C, and adopts XHTML Basic and CSS Mobile Profile as part of WML2.

cHTML is still being used for i-mode devices, but is not as widely used as XHTML Basic and WML. In the future, since cHTML is based in HTML, it could very likely merge with a form of WAP or XHTML.

This book presents all three programming languages, WML (and WMLScript), XHTML Basic, and cHTML. Anyone interested in programming wireless devices today, should really know all three. cHTML and XHTML Basic are very similar, both being based in HTML, and are not as extensive a language as WML and WMLScript. Also, if they all three merge in the near future, you as the reader, will know all three of the components of the merged language.

Chapter 1

Wireless Access Protocol (WAP)

WAP is the standard used for wireless computing. It is basically a protocol, or more specifically, a collection of protocols and standards. WAP is to wireless devices what HTTP (Hyper Text Transfer Protocol) is to Web browsers.

WAP was designed specifically for wireless computing and therefore accommodates the unique and fundamental limitations of wireless computing. Unlike Web browsers running in PCs, in wireless computing the devices have: limited processing power and memory; limited battery power and life; small displays; limited data input and user interaction capabilities; limited bandwidth and connection speeds; and a large number of different types of WAP devices.

The WAP Forum was created in 1997 to oversee and modify as needed the WAP protocols and standards. It started with just a few companies, and within 2 years had over 100 companies in it.

In June 2002, the WAP Forum merged with the Open Mobile Architecture to form OMA – the Open Mobile Alliance. The OMA addresses a larger scope of standards and requirements than the WAP Forum did. The OMA is designed to be a focal point of mobile services specification work, to assist the creation of interoperable mobile services across countries, operators, and mobile terminals. There are currently over 300 companies in the OMA. The OMA basically defines industry wide requirements, common architectural framework, open standards for enabling technologies and end-to-end interoperability. This obviously includes WAP standards.

How WAP Works

Basically, a WAP device – (most are phones, but also PDAs and other handheld wireless devices) initiates a request through the Internet to a WAP Server. A WAP Server responds to the request and sends the requested data back through the Internet to the WAP device.

Wireless Markup Language (WML)

Figure 1.4 shows a WAP device requesting data through the Internet to a WAP Server. First, the WAP use selects content (a document) on their handset that they desire to receive. This creates a request message that is sent to the IP address of the WAP server. The WAP server processes this request and returns the requested document to the IP address assigned to the WAP device. When the WAP device receives the data (document), it can display the document on the screen.

A WAP Server is software, running on a computer, that provides WAP content to requests. WML, XHTML Basic and WMLScript are the languages you write programs in that reside on the WAP Server, and respond to requests from WAP devices.

WAP devices have integrated browsers, called micro-browsers, and a mechanism for user input.

A WAP Gateway, interfaces WAP devices to the Internet. It also can handle requests from WAP devices to HTTP Servers.

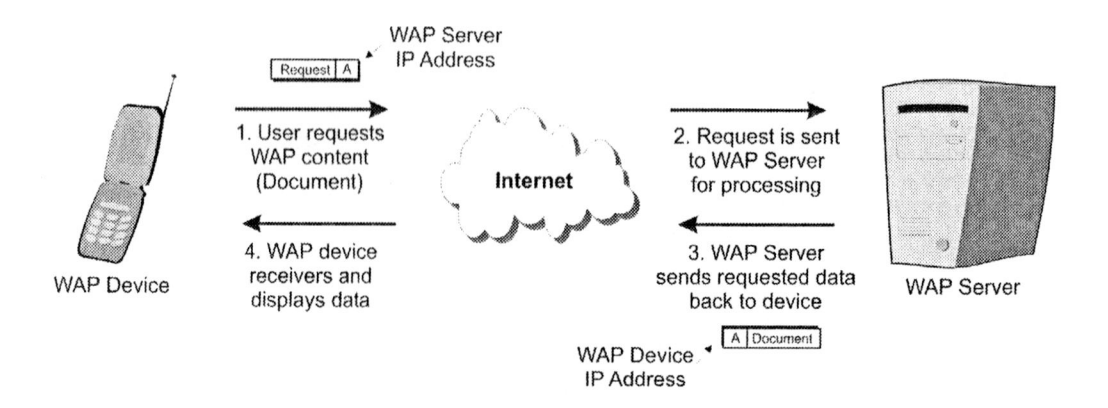

Figure 1.4 WAP Device and WAP Server Operation

Figure 1.5 shows how a WAP Gateway interfaces (adapts) the information between WAP devices to HTTP information sources that are connected the Internet (such as web pages). This example shows that the initial request for content comes from the WAP device that is requested to be connected to a specific HTTP server address (URL address). The connection request message is first routed to a WAP gateway that will be used to interface (adapt) the information between the WAP device and the HTTP information source. The WAP gateway then sends the information request to the desired HTTP server. The HTTP server sends the requested data back to the Internet in a format that may not be compatible with the WAP device (screen size and graphics resolution). The WAP Gateway converts the received HTTP data to a format that can be used by the WAP device and it forwards this information to the WAP device via the Internet.

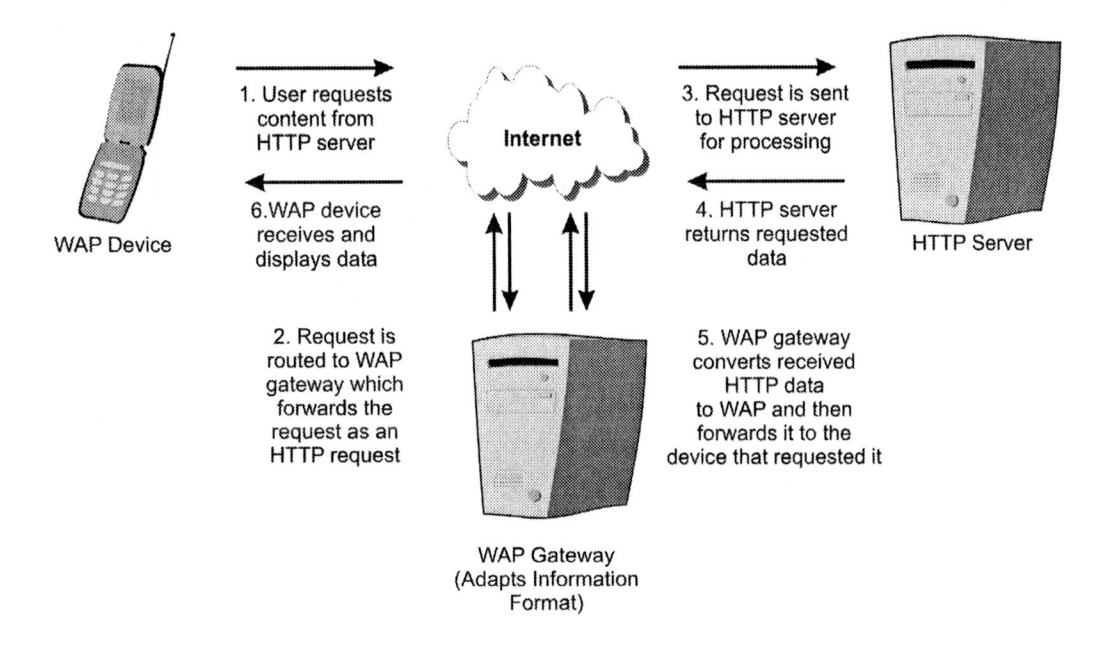

Figure 1.5 WAP Gateway Operation

Wireless Markup Language (WML)

WML – "Wireless Markup Language" is a tag-based language used for describing the structure of documents to be delivered to wireless devices. It is to wireless devices what HTML (Hyper Text Markup Language) is to Web browsers. HTML is the language used to layout web pages to be viewed by a Web browser, whereas WML is the language used to layout "pages" (or cards as they are called in WML) to be viewed in wireless devices. HTTP is the protocol for HTML, and WAP is the protocol for WML.

The browsers in wireless devices are called micro browsers. WML is similar to HTML, but is less forgiving and stricter on syntax than HTML. It is more like a programming language this way than HTML is.

HTML is more concerned with the placement of images and text on the page than WML is. This is because Web pages are large and wireless cards are small, and in WML you have much less control over where certain text and images (small images) show up on the page since all devices can appear differently given the same WML code.

Figure 1.6 demonstrates how a mobile wireless device can access a WML document via the Internet. Initially, the user selects (requests) a stock quote from his stock company to be delivered as a WML document. This request is sent from the wireless device, the cellular tower to the cellular system. The cellular system then forwards this request via the Internet to a stock company's WAP server that has the requested WML code or document stored in its system. After the WAP server has gathered and formatted the requested document, it is then sent via the Internet to the cellular systems that transmits the message to the wireless device.

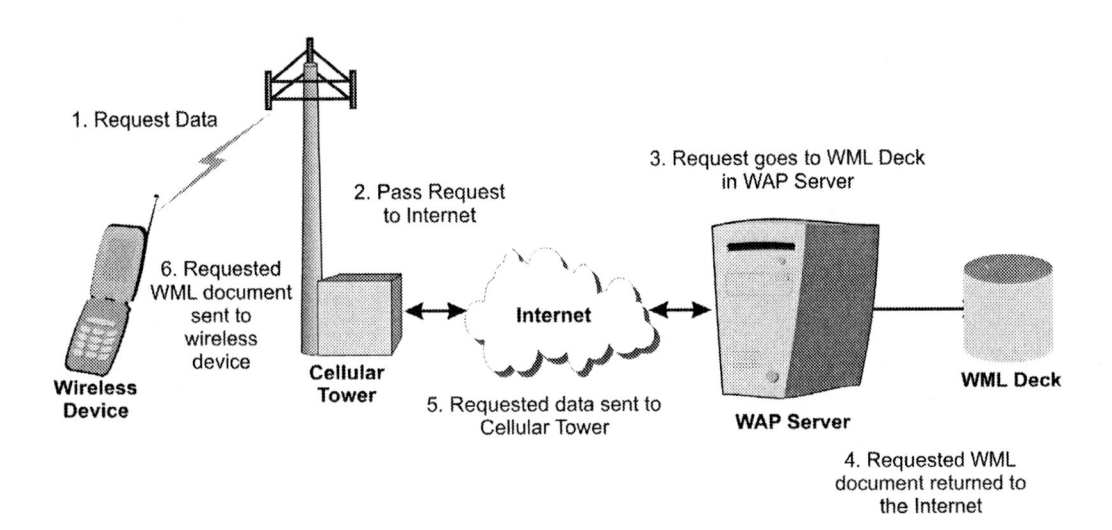

Figure 1.6 WML Document Access Operation

WMLScript

WMLScript is a client-side scripting language that complements WML. It is directly analogous to JavaScript and HTML. WMLScript provides basic programmability that you can use to perform basic text and data manipulation, for WML applications. So, WMLScript allows you to do data manipulation in wireless applications that you can't do within WML. Normally, you call WMLScript routines from within WML programs to manipulate the data inputted to the WML program.

In general, WMLScript is simpler language than JavaScript because JavaScript is a full-featured language that offers almost complete control over the entire user interface on a WWW browser. WMLScript, however, is designed to operate within the context of portable, narrowband devices and is therefore simpler.

Wireless Markup Language (WML)

Table 1.2 shows which language protocols are used for web based wired (low cost high-speed) and wireless (high cost low-speed) communications. This tale shows that the primary transport language for the web is HTTP and the WAP uses WAP protocol. The markup (presentation) language for the web is HTML and WAP uses WML. The primary scripting language (application processes) for the web is Javascript and WAP uses WMLScript.

Feature	Web (Wired)	WAP (Wireless)
Transport	HTTP	WAP
Markup	HTML	WML
Scripting	JavaScript	WMLScript

Table 1.2 Comparing WAP to the Web

Figure 1.7 shows how WMLScript can be used to respond to WML requests for information. This example shows that a user has requested a weather forecast document from his preferred weather information company. This request is sent through the cellular tower to the cellular system that forwards the request to the Internet address selected by the user. The Internet routes the request to a WAP server. The WAP server has a WML deck residing on it. Because this users request involves inputting data (the requested city), a WMLScript function is called from the WML deck to process the inputted data. The WMLScript function then sends the processed data (the temperature and forecast) back to the device via the Internet, through the cellular system, to the display of the wireless device.

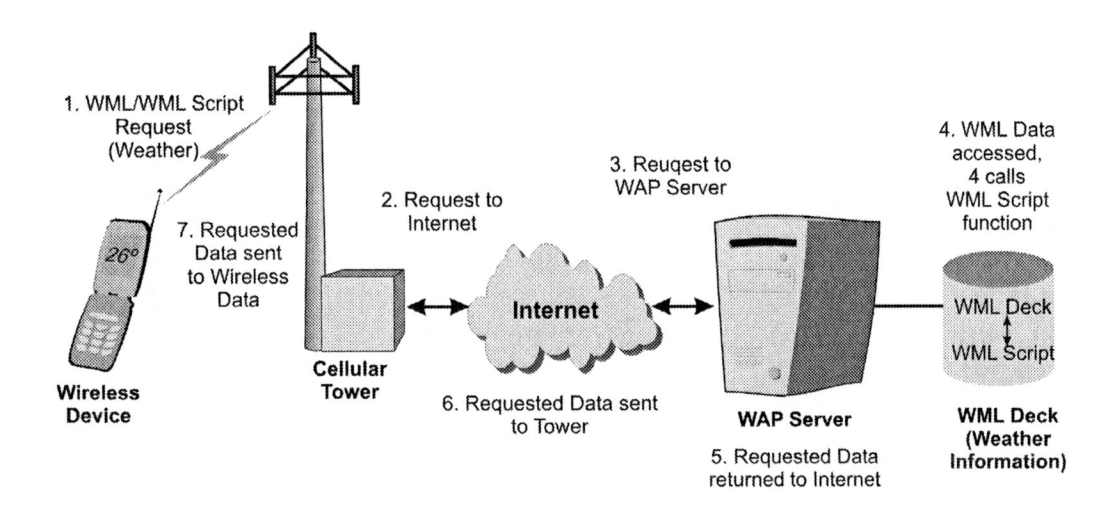

Figure 1.7 WMLScript Operation

XHTML Basic

XHTML (Extensible Hypertext Markup Language) Basic was created by the World Wide Web Consortium (WC3) to provide a common markup language for wireless devices and other small devices with limited memory. WAP 2.0 specifies XHTML Basic as the markup language for wireless devices. WAP 2.0 allows WML to be embedded within XHTML Basic so developers can implement features not supported by XHTML Basic such as soft keys, using WML.

XHTML Basic is similar to WML in that it is derived from XML. XHTML Basic does not include features from XHTML such as frames and nested tables since they are not well suited for wireless clients.

Wireless Markup Language (WML)

Figure 1.8 shows an example of how XHTML Basic operation can be used to process information requests from mobile devices. This example shows that a user has sent movie time request from a Wireless device to their preferred entertainment information provider. This request is routed through the cellular tower and to the cellular system that forwards the request to the Internet using the selected Internet address of the information provider. The Internet routes the request to a WAP server. The WAP server determines that this request is for a document or information that is written in XHTML Basic and stored on the WAP server. The XHTML Basic program is accessed, and the requested data is sent via the Internet through the cellular tower to the wireless device for display. The requested data could be any web site that, for example, has been converted to XHTML Basic from HTML so that it can be displayed correctly on any wireless device.

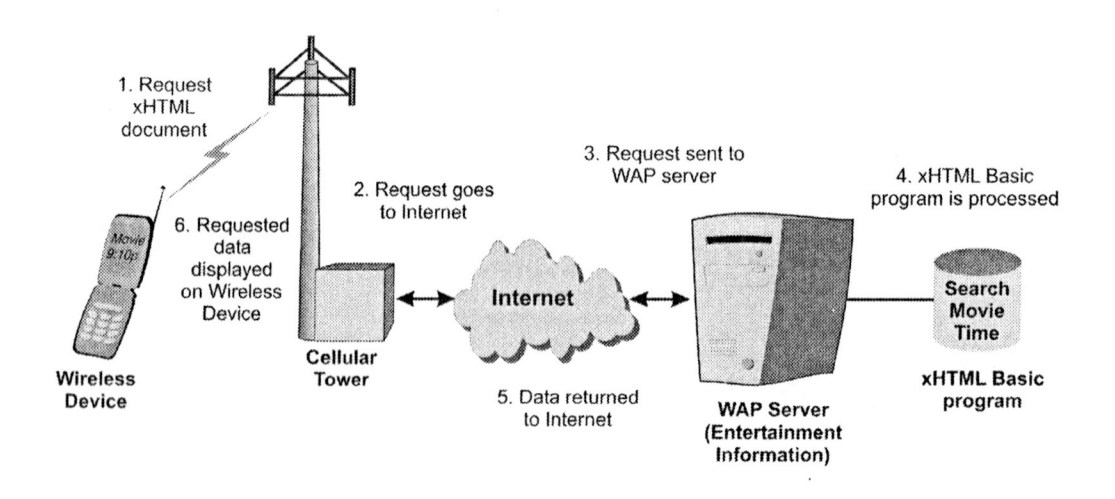

Figure 1.8 XHTML Basic Operation

Chapter 1

cHTML

cHTML was developed by NTT DoCoMo as the wireless language to use in programming the wireless devices that are served by their wireless Internet access service, i-mode. cHTML (Compact Hypertext Markup Language) is a subset of HTML, but cHTML supports fewer character fonts and styles than HTML. cHTML does not support image maps, tables, cookies, frames, JPEG images, style sheets or scripts.

Currently, cHTML is not included in WAP2.0, as is XHTML Basic and WML, but it still is in use for all i-mode wireless devices. It may be incorporated in the future.

Figure 1.9 shows how a user in the i-mode system can access a cHTML program via the Internet using the i-mode ISP. This example shows a user that is requesting a game to be displayed on his wireless device. The request is sent through the cellular system, through the i-Mode ISP network, through the i-Mode server, and is processed by the cHTML program. The program is written in cHTML code to allow the game program to interact efficiently with the wireless device. The cHTML program sends the information through the i-Mode server, through the i-Mode network, through the wireless system, and is displayed on the wireless device.

Wireless Markup Language (WML)

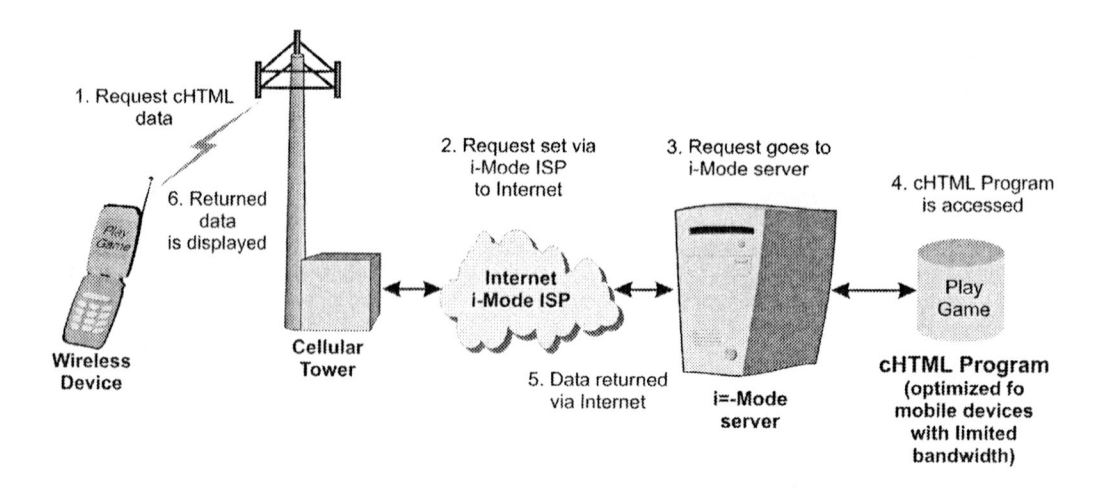

Figure 1.9 cHTML Operation

Summary

- Wireless technology allows individuals and organizations to connect to the Internet and World Wide Web at any time, from almost any location, via wireless devices.

- A cellular network is composed of individual compartments, or cells, which combine to form the communications network. At the center of each cell is a base station that transmits analog or digital signals to and from users within that cell.

- Cellular technology can be considered by first, second, or third generation technology. (1G, 2G, 3G)

- First-generation (1G) technology is the first form of wireless communications networks based on analog signals.

Chapter 1

- 2G systems are designed to use digital-signal transmissions versus 1G analog-signal transmissions.

- 2G technologies include Personal Communications Systems (PCS), Global System for Mobile Communications (GSM), Time Division Multiple Access (TDMA), and Orthogonal Frequency Division Multiplexing (OFDM).

- PCS was introduced in 1995 and uses the digital frequency spectrum. In general, PCS was designed to address the problems in cellular-based networks caused by strains on network capacity limits.

- The Global System for Mobile Communications (GSM) is a PCS digital cellular communications network used in 171 countries.

- Time Division Multiple Access (TDMA) is the foundation for many 2G technologies and will be used in some 2.5G and 3G systems in the future. But, some feel that TDMA could be phased out in favor of other technologies.

- OFDM is based on Frequency Division Multiplexing (FDM). FDM divides frequency channels on the spectrum into smaller channels, or bands, which can be used multiple times.

- Wireless access benefits businesses, employers, employees, and consumers.

- In the future, 3G technologies will help foster Web surfing via wireless devices. These technologies will enable faster connection speeds and the ability to download streaming audio and video.

- Paging is accomplished via pagers that were one of the first widely used devices for wireless communications. Pagers are designed to alert users when new messages arrive and show caller identification information and store messages.

Wireless Markup Language (WML)

- There are currently three main languages being used to program wireless devices, and each one is supported by different companies and consortiums: WML and WMLScript, Xhtml Basic, and cHTML.

- WAP 2.0 specifies WML 2, which encompasses XHTML Basic, and will be backward compatible to WML1.

- WAP is the standard used for wireless computing. It is basically a protocol, or more specifically, a collection of protocols and standards. WAP is to wireless devices what HTTP (Hyper Text Transfer Protocol) is to Web browsers.

- WAP was designed specifically for wireless computing and therefore accommodates the unique and fundamental limitations of wireless computing.

- WML – "Wireless Markup Language" is a tag-based language used for describing the structure of documents to be delivered to wireless devices. It is to wireless devices what HTML (Hyper Text Markup Language) is to Web browsers.

- WMLScript is a client-side scripting language that complements WML. IT is directly analogous to JavaScript and HTML. WMLScript provides basic programmability that you can use to perform basic text and data manipulation, for WML applications.

- XHTML Basic is similar to WML in that it is derived from XML. XHTML Basic does not include features from XHTML such as frames and nested tables since they are not well suited for wireless clients.

- cHTML (Compact Hypertext Markup Language) is a subset of HTML, but cHTML supports fewer character fonts and styles than HTML.

Chapter 1

Questions

1.) What is the transfer of signals called when a user is going between cells?

2.) How does 1G technology differ from 2G technology?

3.) List 4 different technologies that are 2G technologies.

4.) What technology introduced the concept of microcells?

5.) What are two problems with GSM?

6.) Which technology is possibly being phased out in 3G and 4G technologies?

7.) What are some of the ways companies have tried to reduce the inconvenience of entering data in wireless devices?

8.) List the three types of messaging technology used in pagers.

9.) What is the root markup language that WML and XHTML Basic are based on?

10.) Where is cHTML used?

11.) What are some of the limitations of wireless devices that WAP was specifically designed for?

12.) What is a WAP server?

13.) What type of language is WML?

14.) What capabilities does WMLScript provide?

15.) What are two main features that XHTML Basic does not support, and XHTML does?

16.) Is cHTML included in the WAP 2.0 Specification?

Wireless Markup Language (WML)

Chapter 2

Wireless Devices and Operating Systems

Mobile telephones and other wireless communication devices are composed of hardware devices (processors), software operating systems, and applications (such as micro-browsers). These devices are constantly changing and evolving and this chapter presents the current types of hardware that are available and certain widely used operating systems, software platforms, and micro browser applications.

Mobile Devices

Mobile devices include digital assistants (PDAs), digital mobile phones, two-way pagers, mobile data terminals, Pocket PCs, and convergence devices. Mobile devices that include wireless access capability allow users to manage their information while away from their desktop computers. Examples of wireless communication services include mobile telephone service (voice), e-mail messaging (messaging), and multimedia (video). These use of these services allow mobile user applications such as buying airline tickets, ordering groceries, trading stocks, and checking e-mail messages.

Wireless Markup Language (WML)

Mobile Phones

Mobile phones are any handheld devices used to transmit and receive calls from a wireless system. Mobile phones are also called handsets, cell phones, or wireless phones. These devices include antennas for signal transmission, number pads for dialing, speakers and microphones for voice communications, operating systems that control how the phones operate and the computer chips needed to complete wireless communications and other functions. Because mobile phone products vary in size, shape, color, graphics display, and information processing capability, mobile devices can be characterized (grouped) according to their functions and applications.

Flip-phones are compact versions of mobile phones that fold up to minimize their size. Users can also select ring tones already installed on the device or download additional choices from the Web to personalize their phones. Other phone features include phone books, caller ID, and voicemail.

Most mobile phones today have a dual-mode feature that allows them to support analog or digital signals, although digital is more popular.

Some mobile phones have distinguishing features that set them apart. One is Motorola that offers a mobile phone with a display screen using organic electro-luminescent (OEL) technology. OEL allows users to view the display screen in three colors. OEL increases the visibility and clarity of graphics on mobile phones. Another is I-mode mobile phone in Japan that offers full-color screens and high-resolution graphics.

Voice recognition is being offered on some phones for voice-activated dialing which allows the user to dial a number or check messages on a phone using voice commands. In addition to voice recognition, most mobile phones can be used to access the Web, check e-mail, and host games.

Figure 2.1 is a photo image of a sophisticated camera phone from Sony Ericsson that includes a WAP 2.0 browser. The T610 also has cHTML, Multimedia Messaging (MMS), and Enhanced Messaging (EMS), and has Java ™ download capabilities.

Chapter 2

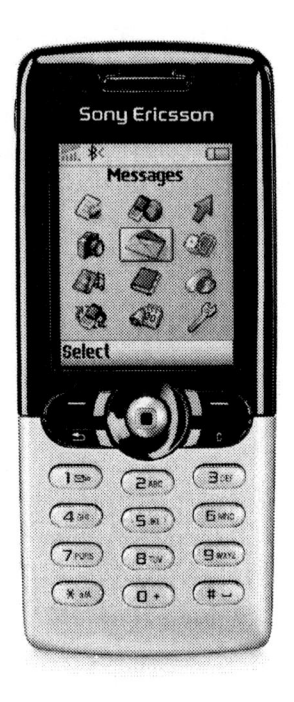

Figure 2.1., T610 Mobile Telephone with WAP 2.0 Browser
Source: Sony Ericsson

Mobile Data Terminals

Mobile data terminals are handheld devices that are most often used for wireless e-mail solutions. These mobile data terminals are available in both pager-sized and palm-sized units and include a keypad, making it easy to type messages quickly. Most wireless handheld devices support the wide-area wireless data network protocols, DataTAC Network and Mobitex Network. Both DataTAC and Mobitex are personal communications services (PCS) networks that enable wide-area wireless data communications. The protocols provide in-building penetration, roaming, messaging services and guaranteed delivery and reliability.

Wireless Markup Language (WML)

Figure 2.2 shows a BlackBerry 7230 Wireless Handheld ™ that provides wireless access to email and the Internet (the BlackBerry 7230 ™ also includes a phone and organizer). This handheld can be used by mobile professionals to gain wireless access to corporate data, such as enterprise email accounts and Intranet information. BlackBerry ® is a totally integrated package that includes software, nationwide/international airtime and choice of wireless handhelds to provide easy access to corporate emailand information on the go.

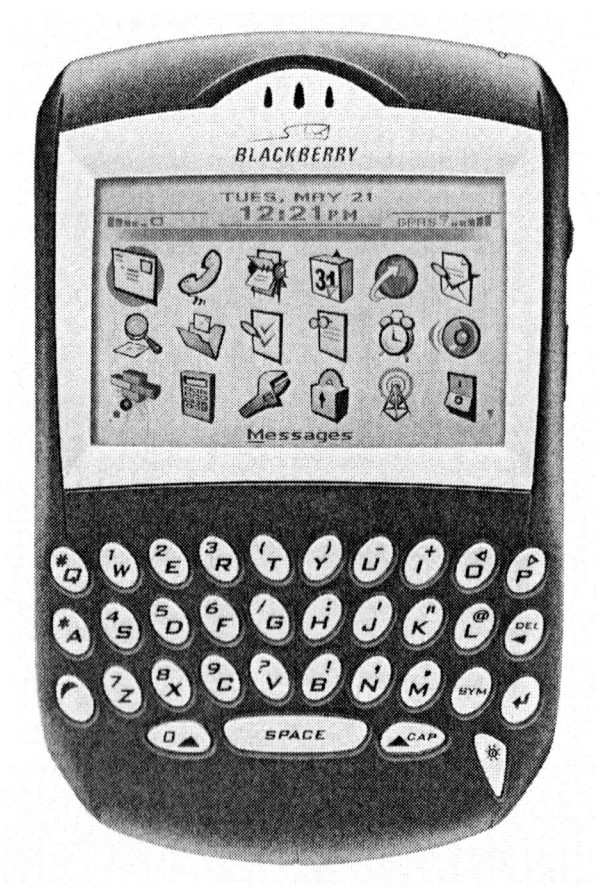

Figure 2.2., Model 7230™ Wireless Handheld with WML
Source: Research In Motion (RIM)

Chapter 2

Two-Way Pagers

A two-way pager is a small radio receiver designed to be carried by a person and to give an audio, visual or tactile indication when activated by the reception of a radio signal containing its specific code. It may also reproduce sounds and/or display messages that were also transmitted. Two-way pagers are interactive as they acknowledge that a message has been received and/or initiate new messages from the pager to the recipient.

Personal Digital Assistants (PDAs)

Personal Digital Assistants (PDAs) are small computing devices that contain it's own software operating system that allows the user to run software processing applications. Personal digital assistants are often used to organize personal activities and may provide access to communication services (such as web browsing and email). PDAs are commonly used to access the Web to check stock quotes, send and receive e-mail, make travel arrangements, find directions and shop online.

PDAs are equipped with infrared technology to communicate and share information with other devices. Future PDAs will use Bluetooth wireless technology to connect to wireless networks and transmit information.

PDAs can be used for numerous business and consumer applications. Employees can access product information and corporate databases from remote locations without using PCs or laptop computers. Doctors and medical staff use PDAs to share patient information and write digital prescriptions. Digital prescriptions have helped reduce the number of errors that occur due to the inability to read a doctors illegible handwriting.

Wireless Markup Language (WML)

Convergence Devices

A convergence device combines multiple functions such as a PDA, mobile phone, GPS receiver and two-way messaging into one device. Convergence devices offer voice capabilities, personal organizing, Web browsing, MP3 audio, short-text messaging, and e-mail.

There are two distinct trends for the development of convergence devices. One trend uses existing designs for mobile phones, which offer voice capabilities, messaging and other features, and combines them with PDA information services. The second trend uses existing PDA models and adds voice capabilities.

Handspring and Mitsubishi have developed hardware that extends the capabilities of their PDA products (the Visor and Pocket PC) to include voice and mobile phone transmissions. The Mitsubishi Trium Mondo phone uses the Windows CE 3.0 operating system found in the Pocket PC and allows users to make phone calls and connect to the wireless Internet.

Figure 2.3 shows a Nokia model 9290 convergence device. The 9290 Communicator combines the functions of a PDA, the power of a laptop, and the freedom of a mobile phone into one communication device. This product is based on a Symbian operating system and GSM wireless phone capability. The Nokia 9290 Communicator has a full HTML Internet browser and includes WAP capability.

There are two distinct trends for the development of convergence devices. One trend uses existing designs for mobile phones, which offer voice capabilities, messaging and other features, and combines them with PDA information services. The second trend uses existing PDA models and adds voice capabilities.

Chapter 2

Figure 2.4., Nokia 9290 Communicator

Source: Nokia

Handspring and Mitsubishi have developed hardware that extends the capabilities of their PDA products (the Visor and Pocket PC) to include voice and mobile phone transmissions. The Mitsubishi Trium Mondo phone uses the Windows CE 3.0 operating system found in the Pocket PC and allows users to make phone calls and connect to the wireless Internet.

Another example of a wireless convergence device is a combination of a wristwatch and pager by Timex and Motorola. This device allows users to receive pages, free news updates, sports and weather reports and offers organizational features.

Wireless Markup Language (WML)

Problems exist that hinder the development and adoption of convergence devices by consumers. Carriers and service providers do not have one standard to regulate the devices' structure and technology support. Therefore, universal communications coverage and services are not available. In addition, PDAs may support one standard structure while mobile phones support another.

Pocket PC

Pocket PCs are basically handheld PCs. They have an operating system and offer limited PC functionality in a handheld device. Microsoft has come out with a series of Windows CE operating systems for Pocket PCs that puts a Windows OS on a handheld PC. They started with Windows CE 1.0 and currently have Windows CE 3.0 available. Pocket PCs with Windows CE compete with Palm OS –based handheld devices in the PDA market, and the Pocket PCs with Windows CE 3.0 are closing the gap with Palm-OS PDA devices.

Although the popularity of Pocket PCs is growing in both the United States and internationally, certain foreign markets have seen more widespread adoption of the Pocket PC. For example, Pocket PC's current European market share is three times its market share in the United States.

Pocket PCs, especially those operating on Windows operating system platforms, support streaming audio and video capabilities that are greatly enhanced by faster network connections. Current manufacturers of Pocket PCs are: Compaq, Casio, and Hewlett-Packard. Microsoft, for the past several years, has emphasized the development of software for mobile computing and wireless devices and has increased its presence in the mobile community. Presently, Microsoft is marketing multiple versions of its software for handheld devices, as well as developing new products designed for the next generation of handheld and mobile devices.

Chapter 2

Figure 2.4 shows the HP iPAQ Pocket PC that uses Windows Mobile 2003 Premium software to enable wireless communications. This devices integrates various types of wireless including 802.11 WLAN, Bluetooth, and IrDA.

Figure 2.5, iPAQ Pocket PC
Source: Hewlett Packard (HP) Company

Wireless Markup Language (WML)

Mobile Operating Systems and Platforms

Mobile devices require an operating system (OS) or a software platform on the hardware device itself, in order to provide the environment to run different software applications and programs on the wireless device. Currently there several different mobile OSs and mobile platforms in use, which often makes adapting an application difficult in that it must be tested on several different OSs and platforms. This section looks at some of the mobile OSs and platforms currently offered.

BREW

Binary Runtime Environment for Wireless (BREW) was developed by Qualcomm in an attempt to standardize runtime environments on different devices. The runtime environment of a device is the platform used to execute applications. The BREW standardized platform allows users to download cell-phone applications over the providers' networks. Prior to BREW, companies developing software for wireless devices had to program specific information based on the device type. BREW simplifies such tasks as upgrading or changing browsers and integrating virtual machines (i.e., self-contained operating systems).

Qualcomm also developed the BREW Software Development Kit (SDK) that runs on Windows NT or higher, for developing and testing applications for mobile devices.

BREW is designed to simplify the integration of cell-phone applications (i.e., e-mail, instant messaging and Internet browsers) and virtual machines in mobile devices. BREW applications are called applets. An applet is a specific type of class and is contained in a module. Modules reside within a program and contain code that performs one or more tasks.

Chapter 2

Palm Operating System (Palm OS)

An operating system (OS) serves as the foundation software for a device. The OS allocates storage, schedules tasks, controls how applications are programmed and run on the device and present a common interface for these applications. The Palm OS controls modes of operation, memory types, applications, performance and other features. There are many versions of the Palm OS, including Palm OS 2.0, 3.0, 3.5, and the latest release Palm OS 4.0. Smart phones, cell or mobile phones, and PDAs can use Palm OS. So, the Palm OS can obviously run on devices other than just those manufactured by Palm.

Palm OS does not support multitasking, like a typical PC OS, but it does have the user interface application shell (UIAS) which manages and prioritizes applications, setting an order in which certain tasks take precedence.

Palm OS 4.0 supports live multimedia, streaming audio and video and cellphone functions. Palm OS 4.0 upgrades existing features, enhancing capabilities and upgrading application support. The new platform also supports wireless Internet capabilities.

EPOC

EPOC is an operating system platform designed for next-generation wireless devices. It is a product of Symbian, Nokia, Motorola, Psion, Ericsson, and Matsushita. EPOC provides computing and communication power for wireless devices comparable to that of today's desktop computers.

EPOC supports both mobile phone and PDA platforms. EPOC technology allows users to send and receive mail, fax messages, and to connect to the wireless Internet at 2.5G speeds. EPOC allows developers to use such programming languages as C++ to develop for platforms like WAP and the Internet. In addition, EPOC will support TCP/IP protocols, GSM, Bluetooth wireless technology and infrared standards. EPOC also provides support for data synchronization with PCs and other devices.

Wireless Markup Language (WML)

Figure 2.5 shows a Sony Ericsson P900 Smartphone that uses the Symbian operating system. This Multimedia Smartphone integrates a wide variety of multimedia services and software applications including voice, video, and data capabilities. It includes multimedia messaging (MMS), a WAP browser, and personal organizer applications.

Figure 2.5 Ericsson P900 Smartphone

Source:Sony Ericsson

Chapter 2

Windows CE

The Windows CE operating system by Microsoft, is designed to work in embedded devices, such as mobile phones, palm-sized PCs and Pocket PCs. Windows CE controls memory allocation, applications, performance and connectivity of handheld and mobile devices.

Microsoft started with Windows CE 1.0 in 1996, then came out with Windows CE 2.0, and the current release of Windows CE are 3.0. Several manufacturers run Windows CE 3.0 on their Pocket PCs, such as: Casio, Compaq, and HP.

Windows CE is modular. Windows CE's nine sections can be broken down into 50 or 60 parts. To employ Windows CE modules in the construction of a new system or the modification of a current one, developers use products like Microsoft Windows CE Platform Builder. This type of software sets guidelines indicating what developers can and cannot change about Windows CE and how it is built into a device or system. These modules include memory and system controls.

Stinger OS

Stinger OS is Microsoft's first version of a smart phone OS based on Windows CE. It is important to know that Windows CE 3.0 is the primary OS for both Pocket PC and Stinger applications; however, Pocket PC and Stinger have different qualities and characteristics to accommodate the specific devices for which they are designed (i.e., PDAs and smart phones). Windows CE 3.0 is the main OS platform, and Pocket PC and Stinger applications are customized around this core OS. Stinger consists of some, but not all, components of Windows CE, making it a smaller subset of the original Windows CE OS. This subset accounts for limitations on smart-phone devices. Stinger does not require all the modules and components that Windows CE supports – many of these are inappropriate for such devices as mobile phones.

Wireless Markup Language (WML)

The market for smart phones (including their OSs) is a new area of wireless development. Stinger-based phones are designed to compete in the market with devices using other operating systems, such as those that use Palm OS, (e.g, Kyocera and Handspring) and Linux-based handhelds.

Stinger currently supports 2G and 2.5G wireless technologies and is expected to support 3G technologies in the near future. In addition, Stinger is compatible with GSM, GSM/GPRS, amd CDMA standards, enabling the operating system to be used on a wide variety of existing and future wireless networks.

MicroBrowsers

Microbrowsers are software applications that run on wireless devices, that are used to display pages or cards, such as those documents written in WML or XHTML. Various microbrowsers are available for different types of devices. For instance, Openwave's Mobile Browser is similar to Microsoft's Internet Explorer, except that the Mobile Browser runs on a wireless device, whereas Internet Explorer runs on a desktop computer.

Figure 2.6 shows the Samsung A600 mobile phone that comes with a WAP version 2.0 microbrowser. This mobile phone includes an integrated camera that can send photographs along with an attached audio message.

Chapter 2

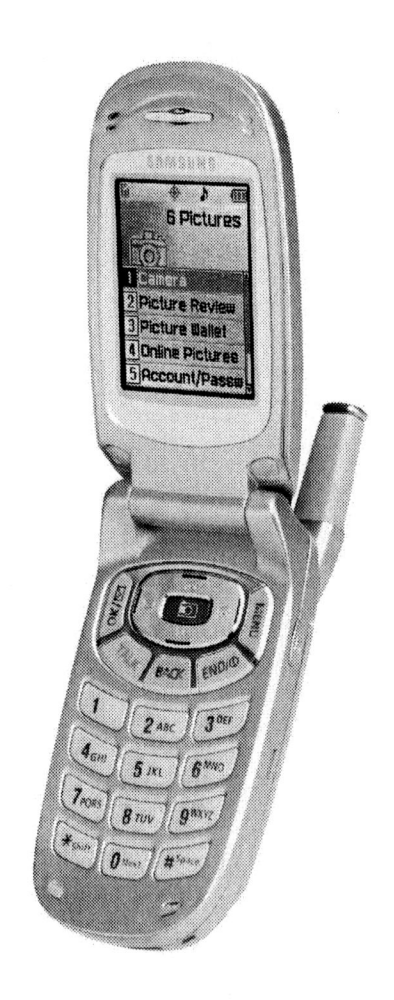

Figure 2.6 Samsung A600 Camera Phone with Microbrowser
Samsung Telecommunication America

Wireless Markup Language (WML)

Summary

- Wireless devices include PDAs, digital mobile phones, two-way pagers, mobile data terminals, Pocket PCs and Convergence Devices.

- Mobile phones are any handheld devices used to transmit and receive calls from a wireless system. Mobile phones are also called handsets, cell phones, or wireless phones.

- Mobile data terminals are handheld devices that are most often used for wireless e-mail solutions.

- Two-way pagers offer two-way interactive communications, allowing users to send and receive text messages.

- PDAs allow users to access the Web to check stock quotes, send and receive e-mail, make travel arrangements, find directions and shop online. PDAs are also organizers that store contact information, appointments and other personal information.

- PDAs, mobile phones, GPS systems, and two-way pagers will be combined into one wireless device called a Convergence Device. One convergence device that already exists is the Kyocera QCP 6035 Smart Phone, which is a Palm handheld, combined with mobile-phone capabilities.

- Pocket PCs are basically handheld PCs. They have an operating system and offer limited PC functionality in a handheld device.

- Mobile devices require an operating system (OS) or a software platform on the hardware device itself, in order to provide the environment to run different software applications and programs on the wireless device.

Chapter 2

- Binary Runtime Environment for Wireless (BREW) was developed by Qualcomm in an attempt to standardize runtime environments on different devices.

- The Palm OS controls modes of operation, memory types, applications, performance and other features. There are many versions of the Palm OS, including Palm OS 2.0, 3.0, 3.5, and the latest release Palm OS 4.0.

- EPOC is an operating system platform designed for next-generation wireless devices. It is a product of Symbian, Nokia, Motorola, Psion, Ericsson, and Matsushita.

- The Windows CE operating system by Microsoft is designed to work in embedded devices, such as mobile phones, palm-sized PCs and Pocket PCs.

- Stinger OS is Microsoft's first version of a smart phone OS based on Windows CE.

- Microbrowsers are software applications that run on wireless devices, that are used to display pages or cards, such as those documents written in WML or XHTML.

Wireless Markup Language (WML)

Questions

1.) List five existing wireless devices.

2.) What are the basic hardware components of a mobile phone?

3.) What is meant by "dual-mode" when referring to mobile phones?

4.) What is the most popular mobile data terminal in use today?

5.) How do two-way pagers differ from Mobile Data Terminals?

6.) List three main functions available on PDAs.

7.) What is a convergence device?

8.) What are some of the problems facing the development and adoption of convergence devices?

9.) Who are some current manufacturers of Pocket PCs?

10.) Why did Qualcomm develop BREW?

11.) What devices can use Palm OS?

12.) What main function of an operating system does Palm OS not support?

13.) What are the three main functions offered by EPOC?

14.) What functions does Windows CE provide?

15.) What is Stinger?

16.) What is a microbrowser?

Chapter 3

Wireless Markup Language (WML) Basics

WML (Wireless Markup Language) is an HTML-type formatting language that has been defined as an XML (Extensible Markup Language) document type. WML is not a procedural language such as C or Pascal, since it doesn't indicate how a program is executed. Rather, it is a markup language that identifies the elements of a document so that a wireless device can display that document. WML is ideal for use with wireless devices because it is designed to accommodate their limited memory capacities and small display screens. WML uses formatting commands to describe formatting, location, hyperlinks, images and forms in the text. WAP is the protocol for WML programming, and the WAP Forum developed WML

Basic Structure

WML is set up in the form of cards and decks. A WML document is called a deck, and it consists of one or more cards. A deck is analogous to an HTML page and is identified by a URL address. Cards are basically a screen of text or a screen of user interaction. And, a deck is composed of one or more cards. A deck is the smallest WML unit a server can send to a browser program.

WML documents delimit information with start and end tags. A start tag for a WML document is <wml> and the end tag consists of a name preceeded

Wireless Markup Language (WML)

with a forward slash (/) within the angle brackets, (e.g. </wml>). The code within the <wml> start tag and the </wml> end tag is considered a deck.

All WML documents must be preceded by the same three lines of code that define the versions of WML and XML being used and are needed for the browser to be able to understand the WML code. Example 3.1 shows these three lines.

```
<?xml version="1.0"?>
<!DOCTYPE wml PUBLIC "-//WAPFORUM//DTD WML 1.1//EN"
"http://www.wapforum.org/DTD/wml_1.1.xml">
```

Example 3.1

All WML documents must contain these three lines of code.

The <wml> ...</wml> command defines the deck and locks all information and cards inside the deck. Example 3.2 shows a simple WML deck containing a single card that displays "Hello!" on the screen. How this appears on the screen of a Nokia WAP Toolkit is shown in Figure 3.1.

```
<?xml version="1.0"?>
<!DOCTYPE wml PUBLIC "-//WAPFORUM//DTD WML 1.1//EN"
"http://www.wapforum.org/DTD/wml_1.1.xml">

<wml>

 <card>
<p>
Hello!
</p>
</card>

</wml>
```

Example 3.2

Figure 3.1 WML Hello Screen

Example 3.2 shows the use of the card command <card> and the paragraph command <p>. All commands in WML need to be "ended" with a </command>. These commands will be discussed in more detail later in this chapter. Also, most commands have attributes that are listed with the command within the <..> 's.

WML encoding does not allow the expression of all symbols in a document. In such cases character entities can be used. WML supports both alphanumeric (e.g. < is represented by <) and numeric (e.g. { is represented by a decimal character entity { or in hex {).

Some Typical Character Entities

Quotation Mark	"	&quot;	&#34;
Hash mark	#		&#35;
Dollar	$		&#36;
Percent	%		&#37;
Ampersand	&	&amp;	&#38;
Apostrophe	'	&apos;	&#39;
Smaller than	<	&lt;	&#60;
Greater than	>	&gt;	&#62;

Table 3.1

Wireless Markup Language (WML)

Inserting Comments

Comments in WML are similar to those in HTML. A single comment starts with <!-- and it ends with -->. The browser ignores text within the comment tags, and the text is never shown in documents or elsewhere on the browser's display. The comments are only for readability of the code, or for future markers in the code. Comments can be split over several rows, but cannot be nested. Example 3.3 shows valid comments in WML.

```
<!-- this is a valid comment -->

<!-- comments may also span
 several rows as long as they
 are not nested. -- >
```

Example 3.3

Variables

WML programming allows for the definition of variables, unlike HTML. Variables can be used between cards in a deck, but most often they are used in interacting with WMLScript functions or when passed to a server to be processed by a CGI program.

Variables can be used in the WML document body text, and are defined using the <setvar/> command with the required attributes **name** and **value**. The command can be used to transfer information between cards when navigating, in which case the command is usually placed within a<go>...</go> , <prev>..</prev> or a <refresh>..</refresh> command, depending on the task. The **name** attribute of the <setvar/> command defines the identifier or name of the variable, and the **value** attribute sets the value of the variable. Variable names are case-sensitive. When referring to a variable name in the text, (except within commands), its identifier or name should begin with a dollar ($) sign. The identifier must start with a letter or an underscore (_) and the other characters can be letters, numbers or underscores. WML code requires the variable to be enclosed in parentheses (), only if the variable identifier cannot otherwise be distinguished from the body text.

Example 3.4 shows some examples of valid variables.

Valid Variable Identifiers

$var1 is a legal variable.

$(var1) is also a legal variable

$_variable1 is a legal variable

$(_variable1) is the same exact variable as above

$variable and **$Variable** are both legal but are different variables.

Example 3.4

Example 3.5 shows some examples of invalid variable identifiers.

Invalid Variable Identifiers

$$ is a dollar sign, not a variable

$11eleven is not legal

$-mem is not legal

Example 3.5

Wireless Markup Language (WML)

```
<?xml version="1.0"?>
<!DOCTYPE wml PUBLIC "-//WAPFORUM//DTD WML 1.1//EN"
"http://www.wapforum.org/DTD/wml_1.1.xml">


<wml>
 <card id="card1" title="Variable Example">

<do type="accept" label="Initialize">
 <refresh>
 <setvar name="name" value="Bill"/>
 </refresh>
 </do>
 </card>


</wml>
```

Example 3.6

In example 3.6 the variable "name" is assigned a value of "Bill" in a <setvar/> command when the button labeled "Initialize" is pushed. Figure 3.2 shows what the screen looks like with the code from example 3.6.

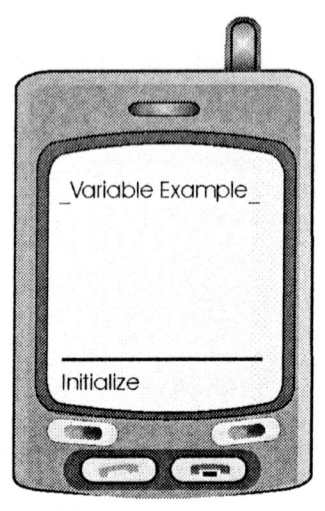

Figure 3.2 Variable Example

Example 3.7 shows the assigning of an initial value of "William" to the variable "name" in the command <input> which is requesting user input. So, once executed, the variable "name" will have an initial value of "William" assigned to it, but once the user inputs a value, that will become the value of the "name" variable.

```
<?xml version="1.0"?>

<!DOCTYPE wml PUBLIC "-//WAPFORUM//DTD WML 1.1//EN"

"http://www.wapforum.org/DTD/wml_1.1.xml">


<wml>

<card id="card1" title="Variable">
<p>
Your Name: <br/>
<input title="name" name="name" value="William"/>
</p>

</card>


</wml>
```
Example 3.7

Figure 3.3 shows what the first screen looks like from example 3.7 and it also shows the second screen that appears after the Edit button is pushed. It is requesting input from the user, but shows an initial value of William in the input box.

Wireless Markup Language (WML)

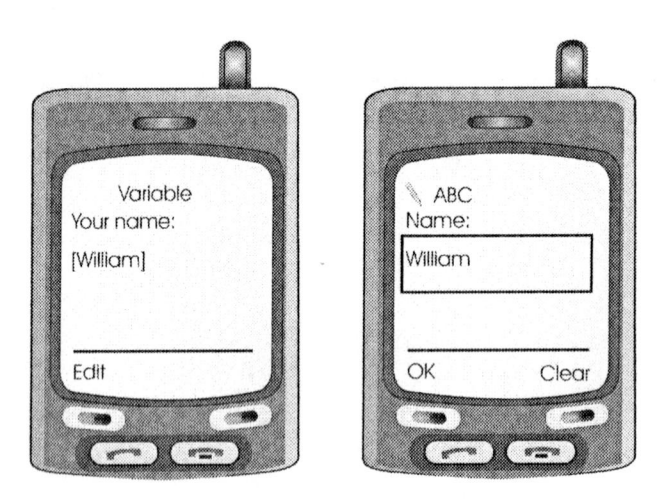

Figure 3.3 Variable Example 2

The <input> command will be discussed in detail later in this chapter, but it can also be used without assigning any initial value. In those cases, the input box as seen in figure 3.3 will be empty, waiting for the user's input.

Type Conversion of Variables

Type conversion of variables or substitution modifiers for variables, can be used explicitly on a variable. The form is $(variable:conversion). The conversion can be **escape, unesc**, or **noesc**. These all define the type conversion of the variable value during value assignment. This means that special symbols are coded before the value of the variable is replaced with a new value if the name of the variable is followed by **escape**. Conversely, **noesc** after the variable name prevent special symbols from being coded. If **unesc** is used, all the special symbols that may have been coded during value assignment will be decoded or converted back to the original symbols. An example of this is URL escapement, which is when a browser takes any special characters in a URL and replaces them with the ASCII value in hex, preceded by a %. For example a question mark would be replaced with a %3F. So, if your program passes variables between decks or to a script, and special characters get inserted in the value assignment, you can specify whether or not to remove these characters, or not escape them at all, or leave them escaped.

Text Formatting

The way WML does text formatting is similar to HTML. Any bold and italicized source text formatting is not displayed in the browser. In order to get bold or italics, tags must be used. .. tags are put around text to make it bold. <i>...</i> for italics, ... and ... for emphasis and highlighting, <u>...</u> for underlining, <big>...</big> for increasing font size and <small>...</small> for decreasing the font size. This WML formatting is similar to HTML in form, but in WML not all browsers interpret all formatting tags the same. So, you have less control over exactly how your formatted text will appear in different browsers. Example 3.8 shows how these different formatting tags are used for different effects.

```
<?xml version="1.0"?>
<!DOCTYPE wml PUBLIC "-//WAPFORUM//DTD WML 1.1//EN"
"http://www.wapforum.org/DTD/wml_1.1.xml">


<wml>
 <card id="card2" title="Formatting">

<p>

<em> Usually </em> <strong> there is</strong> <i> no </i>
<b> reason</b> <u> to</u> <big> emphasize </big>
<small> text </small>
</p>
</card>

</wml>
```

Example 3.8

Wireless Markup Language (WML)

Figure 3.4 shows the corresponding screens for the code in example 3.8.

Figure 3.4 Formatting

In WML as in HTML, the <br/> command causes a line break in the display. Since the command does not require an end tag, the command is followed by a slash inside the brackets. (This is different from HTML) When the browser sees a <br/> command, it ends the current line and starts a new line. Example 3.9 shows the use of the <br/> command. Figure 3.5 shows the resulting screen. Take note that consecutive breaks are usually ignored by the browser. (In some PDA's multiple break commands will cause multiple line breaks to occur.)

Chapter 3

```
<?xml version="1.0"?>
<!DOCTYPE wml PUBLIC "-//WAPFORUM//DTD WML 1.1//EN"
"http://www.wapforum.org/DTD/wml_1.1.xml">


<wml>
 <card id="card3" title="Line Break">

<p>

First new line <br/>

Second new line <br/><br/>

Third line
</p>
</card>

</wml>
```

Example 3.9

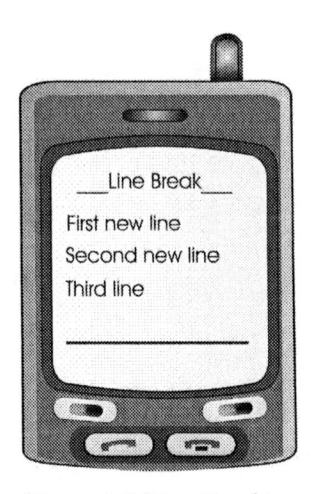

Figure 3.5 Line Break

The <p>...</p> command is used in WML as it is in HTML to define a paragraph. By default, the paragraph is aligned left but the align attribute can be used to align the paragraph to the right or the center. The allowed values for the align attribute on the paragraph command are **left, center**, and **right.** The mode attribute of the <p> can have two values – **wrap** and **nowrap**. If the mode is set to wrap (the default case), then the browser will be able to break the text when executing a line break. If the attribute is set to nowrap, the browser will try to show the text without line breaks. This can often lead to only partially displayed text due to the small screens of cell phones. Example 3.10 shows some examples of using the alignment attribute of the <p> command, and figure 3.6 shows the screens for the code in example 3.10.

```
<?xml version="1.0"?>
<!DOCTYPE wml PUBLIC "-//WAPFORUM//DTD WML 1.1//EN"
"http://www.wapforum.org/DTD/wml_1.1.xml">


<wml>
<card id="card4" title="Paragraph Align">

<p>
The paragraph aligned to left (default),
</p>
<p align="right">
The paragraph aligned to right.
</p>

<p align="center">
The Paragraph aligned to center.
</p>

<p mode="wrap">
Lines can be wrapped (default),
Browser handles new lines.
</p>
<p mode="nowrap">
The programmer should take care of the new lines,
because lines are wrapped.
</p>
</card>
</wml>
```

Example 3.10

Wireless Markup Language (WML)

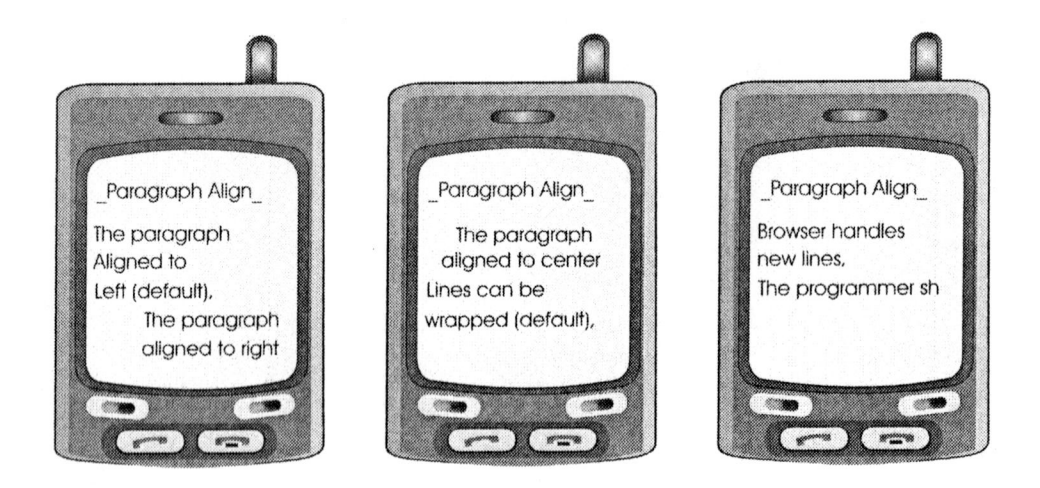

Figure 3.6 Paragraphs

Basic Commands

All commands in WML have the form of <command>....</command>. Commands need to be ended with a </command>, or else be one of the commands that can be "ended" within the one set of command brackets, such as <br/>. Also, commands can have many attributes that go within the first command's set of brackets, e.g. <command attribute1="value1" attribute2="value2">. These attributes define the command or set up different conditions or settings of the command. Some commands have required attributes that must be there always, as well as optional attributes. Other commands have only optional attributes or in some cases, no attributes. Appendix I lists all the WML commands and their associated attributes.

Card

The card command <card>..</card> defines a card in a WML deck. Each card in a deck must start and end with a card command. The card command has no required attributes, but has many optional ones. The **id** attribute specifies the name associated with that card, which can be used when transferring between cards in a deck. If the **id** of a card is "card1", then when another card transfers to this card, or refers to it, it uses "#card1". This allows hyperlinks to be used between cards using the id of a card as a relative address.

The **class** attribute is used in order to link the card to one or more groups. So, cards can have individual id's but be put in the same group.

The **title** attribute specifies the name that shows up at the top of the display when this card is being executed. This name the user sees and should define what the card is or does.

Other attributes, **onenterforward** and **onenterbackward**, can be used to specify a URL that the browser moves to when this card is entered going forward or entered going backward. The URL can also be a relative card address within the same deck.

The **ontimer** attribute will be discussed more in the next chapter under timers, but it specifies a URL for the browser to go to when the timer command times out.

The **newcontext** attribute can have a value of "true" or "false". When **newcontext** is set to "true", the browser will clear its history field, and variables from previous cards will be removed from the memory. If the **newcontext** attribute is set to "false", the variables from previous cards will not be removed from memory and the browser will not clear its history field.

The **ordered** attribute may be either "true" or "false". If the **ordered** attribute is set to "true", the card will be placed in linear order according to the WML code order. This feature can be used when cards must be filled out for a form in a specific order. If set to "false", then the browser can determine the order of the presentation elements of the WAP display. It should be noted that the Nokia WAP Toolkit does not support the use of the **ordered** attribute. This makes the broader point, that due to the large number of different types of wireless devices and their browsers, not all browsers or platforms support all WML commands or attributes.

Wireless Markup Language (WML)

Example 3.11 shows the use of the **id, title** and **newcontext** attributes of the card command.

```
<?xml version="1.0"?>
<!DOCTYPE wml PUBLIC "-//WAPFORUM//DTD WML 1.1//EN"
"http://www.wapforum.org/DTD/wml_1.1.xml">



<wml>

 <card id="card1" title="Card Example" newcontext="true">
<p>
This shows the use <br/>
of certain card attributes.
</p>
</card>

</wml>
```

Example 3.11

Figure 3.7 shows how example 3.11 looks on a mobile telephone display.

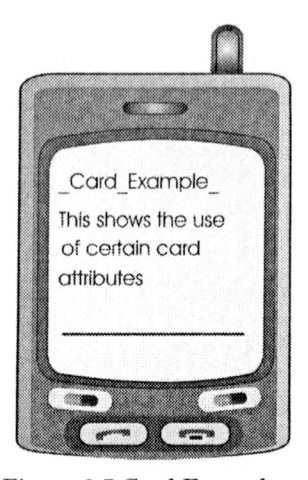

Figure 3.7 Card Example

Do

The do command <do>..</do> starts card tasks. The <do> element appears as a button or a mobile phone key, with a simple action-triggered function. The **type** attribute specifies the type of action that is to take place and is a required attribute. The type attribute must have one of the allowed values listed in table 3.2.

Type Values	Type Attributes Descriptions
accept	Affirms acknowledgement
prev	Provides backward navigation through the history stack
help	Requests for help(can be context-sensitive)
reset	Clears and resets the current state
options	Presents an option to the user that she can elect to use or ignore
delete	Deletes the current choice or item
unknown	Sets a generic <do> tag which is equivalent to an empty string

Table 3.2

The **label** attribute is used to specify the text that is visible to the user and starts the task related to the command. The value of the **label** attribute can be no longer than six characters, since most browsers display the command-related text behind the "options" button, and hence restrict it to as little as six characters.

The **optional** attribute value may be either false (default) or true. The true value tells the browser that it can exclude the element from display due to lack of space.

Wireless Markup Language (WML)

The **name** attribute is used to give a unique name to this do event. This can be important if you have multiple do tags in the same card or deck.

Example 3.12 shows a deck using the do command. The type specified is "accept", and therefore when the button labeled "Forward" is pushed, the action specified within the do command is done. This command is a go com-

```
<?xml version="1.0"?>
<!DOCTYPE wml PUBLIC "-//WAPFORUM//DTD WML 1.1//EN"
"http://www.wapforum.org/DTD/wml_1.1.xml">



<wml>

 <card id="card1" title="Do Example">

<do type="accept" label="Forward">
<go href="#card2"/>
</do>

<p>
Choose <b> Forward </b> to access the next card.
</p>
</card>

<card id="card2" title="Do Example2">
<p>
This is card 2
</p>
</card>


</wml>
```

Example 3.12

mand, and in example 3.12 it goes to card2 when pushed. Figure 3.8 shows the first screen when card1 is executed, as well as the second screen when card2 is executed.

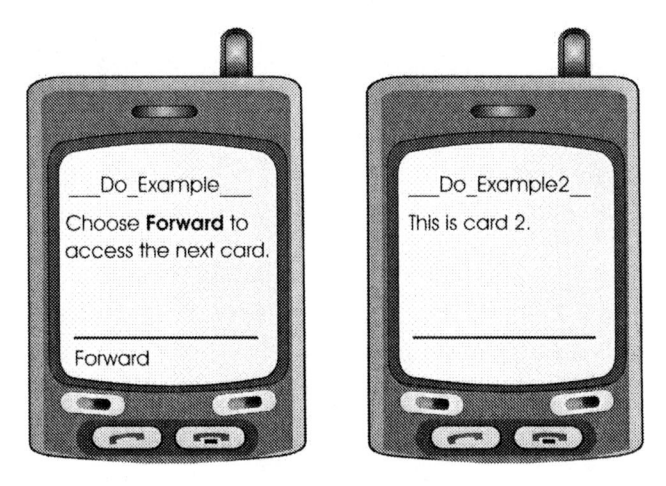

Figure 3.8 Do Example

This example shows the accept attribute which is the most common attribute used in the <do> command.

Go

The <go> command can be used without an end tag (<go/>) or with an end tag <go>...</go>. In the second format, a <postfield/> or a <setvar/> command can be used inside the <go> tags to specify a value for the variable. One or several <postfield/> can be used to define name-value pairs when sending information from a browser to a server. The format of the <go/> command without an end tag is recommended for use if no other commands are defined within the <go> command.

The <go> command defines a transfer to a new URL address – either a deck or a card – as a result of an event. The required **href** attribute is given a relative or an absolute address. The absolute address always contains the complete address, starting with the protocol specification and server name, and ending with subdirectories and the document name (http://wap.acta.fi/wap/document.wml). In a relative address the called document is specified relative to the referring document in the directory hierarchy (subdirectory/document2.wml). If the called document is located on the same server and in the same directory, then a file name alone is sufficient (document3.wml). Also, if the transfer is to another card (say card1) in the same deck, then #card1 is used as the relative address.

If the value of the **sendreferer** attribute of the <go> command is true, the browser will include the URL address of the referring deck in the HTTP request for a new document.

The **method** attribute can have the value of post or get. These values refer to the HTTP method of sending data to the server. The default value is get, but the post method is recommended in CGI programs for security reasons. Also, it may not be possible to send long strings with the get method and so this can't always be used for filling in forms.

The **accept-charset** attribute is used to define the character set that the server should use when processing entries. It is recommended that this attribute always be used when referring to a document or executable program on a server.

Example 3.13 shows the <go> command being used to go to a "help" card in the same deck as the first card. Figure 3.9 shows the displays for example 3.13. The second screen occurs when the user pushes the options button, and the third screen occurs when the select button is pushed.

Chapter 3

```xml
<?xml version="1.0"?>
<!DOCTYPE wml PUBLIC "-//WAPFORUM//DTD WML 1.1//EN"
"http://www.wapforum.org/DTD/wml_1.1.xml">

<wml>

 <card id="card1" title="Go Example">
<do type="help" label="Help">
<go href="#help"/>
</do>
</card>
<card id="help" title="Go Help">

<p>
You can use Go command in a variety of ways!
</p>
</card>

</wml>
```

Example 3.13

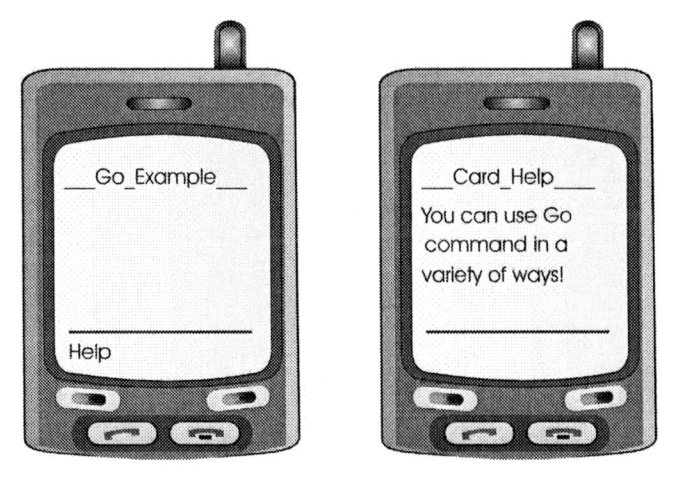

Figure 3.9 Go Example

Wireless Markup Language (WML)

Input

The <input> command is used to create a text field into which the user can enter text. Basically, WML allows for two basic ways for the user to input data. In a text field using the <input> command or in selection lists using the <select> command. The <select> command is discussed in detail in Chapter 4. Data input by the user in one of these two methods can be passed on to a WMLScript function for processing, or to a CGI program or a Java servlet at the server.

The only required attribute for the <input> command is the **name** attribute. This defines the identifier or name for the variable that can be used when handling the string entered into the text field.

The **type** attribute must be set to either text (default) or password. If the **type** attribute is set to text, the input is seen by the user as entered. If the **type** attribute is set to password, only asterisks (*) appear for each character entered.

The **value** attribute defines the default value, which is displayed in the field before the user enters his text. Example 3.14 shows the use of the <input> command and the name, type, and value attributes.

Chapter 3

```
<?xml version="1.0"?>
<!DOCTYPE wml PUBLIC "-//WAPFORUM//DTD WML 1.1//EN"
"http://www.wapforum.org/DTD/wml_1.1.xml">

<wml>

 <card id="card1" title="Text Field">

<p>
Hello!<br/>
Your name:
<input type="text" name="name" value="George"/>
</p>

</card>

</wml>
```

Example 3.14

Figure 3.10 shows the first screen seen when the deck in example 3.14 is executed. When the user pushes the edit button, the second screen appears. If the value attribute is not used, the input box on the second screen appears empty, waiting for user input.

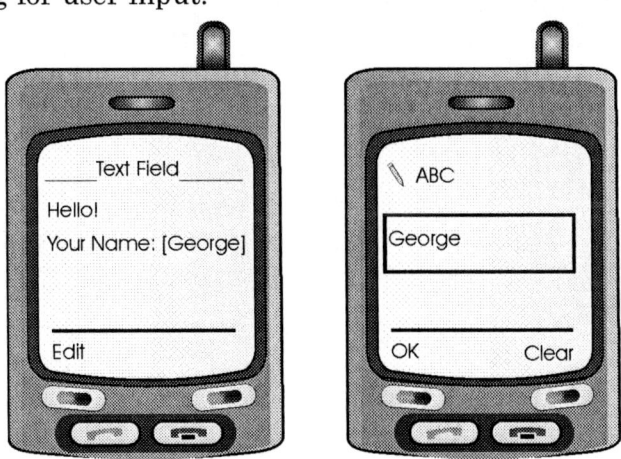

Figure 3.10 Text Field

Wireless Markup Language (WML)

One of the other attributes of the <input> command is the **format** attribute. This attribute specifies, if used, the specific format that the user must use when inputting in terms of characters, numbers, and the number and layout of each. Table 3.3 lists the allowable values for the **format** attribute and what each value represents.

Format Attribute Value	What the value allows
A	capital letters or punctuation marks
a	lower case letters and punctuation marks
N	any digit
X	all upper case letters (no punctuation marks)
x	lower case letters(no punctuation marks)
M	any character, the browser may assume its upper case
m	any character, the browser may assume its lower case

Table 3.3

In the format attribute 3N represents exactly 3 numbers. The wild card (*) represents any number of characters or numbers. *N represents any number of numbers. Also, the "\" is used to specify a character that follows, and the browser will insert this special character in the format where shown. For example, NNN/-NN/-NNNN, represents a persons social security number separated by dashes. As the user enters the first three digits, the browser will automatically insert a dash, and after the user enters two digits, a second dash is inserted, then the user enters the last 4 digits. If the user does not enter the data in this exact format, the browser will not leave the input screen.

Example 3.15 shows how to create a text field, with a specific layout for a social security number.

Chapter 3

```
<?xml version="1.0"?>
<!DOCTYPE wml PUBLIC "-//WAPFORUM//DTD WML 1.1//EN"
"http://www.wapforum.org/DTD/wml_1.1.xml">

<wml>

 <card id="card1" title="Social Security">

<p>
Enter your social security number:
<input type="text" name="ssn" format="NNN\-NN\-NNNN"/>
</p>

</card>

</wml>
```
Example 3.15

Figure 3.11 shows the screens that occur with example 3.15.

Figure 3.11 Input Example2

The **emptyok** attribute for the <input> command can be either true or false. If it is set to false, the browser alerts the user if the field is empty. If set to true, which is the default value, the browser does not alert the user if the field is empty.

The **size** attribute defines the display length of the text field in the browser, and the **maxlength** attribute defines the maximum number of characters which can be inputted.

The **title** attribute is used to define the text for the entry field displayed on the browser screen. This is the name or title that shows up above the text box on the input screen. However, some browsers can choose not to display this title.

The **tabindex** attribute refers to the element activation order; fields with smaller values are activated before ones with greater values when moving between fields.

Anchor

The <anchor>...</anchor> command is used to define a hyperlink. You cannot nest links in WML. Links can be placed anywhere in the text, except inside an <option> command. A link defined by an <anchor> has to be connected to a task that is triggered when the link is activated. The task must be **go**, **prev**, or **refresh**. The **title** attribute for the <anchor> command defines the text that identifies the link, and the maximum recommended length is six characters. The <anchor> command can be used to pass variables. Example 3.16 shows the use of the <anchor> command and also defines a variable to be passed on, named "variable". Figure 3.12 shows the screens associated with example 3.16.

```
<?wml version="1.0"?>
<!DOCTYPE wml PUBLIC "-//WAPFORUM//DTD WML 1.1//EN"
"http://www.wapforum.org/DTD/wml_1.1.wml">

<wml>
<card id="myCard" title="Anchor Example">

<p>

There can be a
<anchor title="LINK"> link
 <go href="Inkex.wml">
 <!—a variable and a value can be defined inside the anchor -->

 <setvar name="variable" value="value"/>
 </go>
</anchor>
between normal text.
</p>

</card>
</wml>
```
Example 3.16

Wireless Markup Language (WML)

Figure 3.12 Anchor Example

The <a>...</a> command is like the command in HTML, and can also be used to create a link. The <a>...</a> command, however, cannot be used to pass variables. The required attribute **href** of the <a> command specifies the link target, and can be a relative or absolute URL address. The **title** attribute can be used to specify the text that identifies the link. If no title is specified, the browser will display a default text "Link" when the link is activated. Nested links are not allowed, and images may also be the link instead of text.

Example 3.17 shows examples of both relative links and absolute links. Figure 3.13 shows the associated screens for this example.

```
<?wml version="1.0"?>
<!DOCTYPE wml PUBLIC "-//WAPFORUM//DTD WML 1.1/
"http://www.wapforum.org/DTD/wml_1.1.wml">

<wml>
<card id="card1" title="Anchors">

<p>

This anchor points to
<a href="#card2" title="Next"> another card </a> in the
same document.<br/>

This anchor points to
<a href="document1.wml" title="Document"> another document </a>
in the same server.<br/>

This anchor points to a
<a href="http://wap.sonera.net" title="Ale"> document </a>
in another server.

</p>
</card>

<card id="card2" title="Anchor2">
<p>
This is card2.
</p>

</card>
</wml>
```

Example 3.17

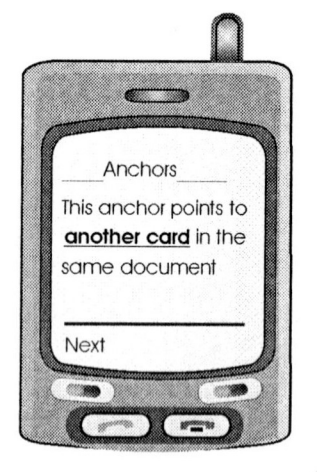

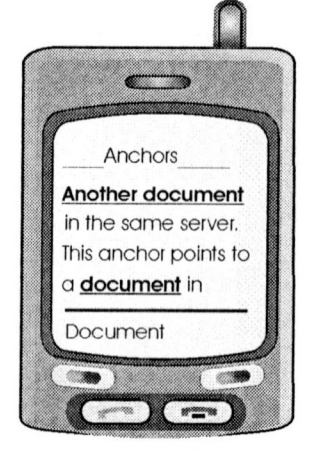

Figure 3.13 Anchor Example 2

Table

Tables in WML are created the same way they are in HTML. The <table>...</table> command defines the table, the <tr>...</tr> command defines the table rows, and the <td>...</td> command defines the separate cells. The table cells may contain text, line breaks, links, or images. Tables may not be nested in WML, and the browser makes the final decision about the presentation of the tables, as well as the formatting of the page.

The **column** attribute is a required attribute for the <table> command. It specifies the number of columns in the table and must be a number greater than zero. Rows are created by using the <tr> command, and the number of rows is not specified in the <table> command like the number of columns is. Example 3.18 shows how to create a table of two rows and two columns. The first column contains a person's name, and the second column that person's phone number. Figure 3.14 shows the screen for example 3.18.

Chapter 3

```
<?xml version="1.0"?>
<!DOCTYPE wml PUBLIC "-//WAPFORUM//DTD WML 1.1//EN"
"http://www.wapforum.org/DTD/wml_1.1.wml">
<wml>
<card id="card1" title="Phone No.s">
 <p>
<table columns="2">

<tr><td>Fred </td>
<td> 555-1234 </td></tr>
<tr><td>Mike </td>
<td>555-9876</td></tr>

</table>
</p>
</card>
</wml>
```

Example 3.18

Figure 3.14 Table Example

Wireless Markup Language (WML)

The **align** attribute for the <table> command specifies the alignment within each column to be **L**-left, **C**-center, or **R**-right. If you have a table with three columns, and the **align** attribute is set to LCR, then the first column will be left adjusted, the center column will be center adjusted, and the third column will be right adjusted.

It is also possible to give the table a name by using the **title** attribute. Example 3.19 shows a three column table with the align attribute and the title attribute being used. Figure 3.15 shows the associated screen. You will note that the title is not displayed in Figure 3.15. Not all browsers execute all attributes.

```
<?xml version="1.0"?>
<!DOCTYPE wml PUBLIC "-//WAPFORUM//DTD WML 1.1//EN"
"http://www.wapforum.org/DTD/wml_1.1.wml">
<wml>
<card id="card3" title="Table">

<p>
 <table columns="3" align="LCR" title="Numbers">

<tr><td>one </td>
<td>two </td>
<td>three </td></tr>
<tr><td>4 </td>
<td>5 </td>
<td>6 </td></tr>

</table>

</p>
</card>
</wml>
```

Example 3.19

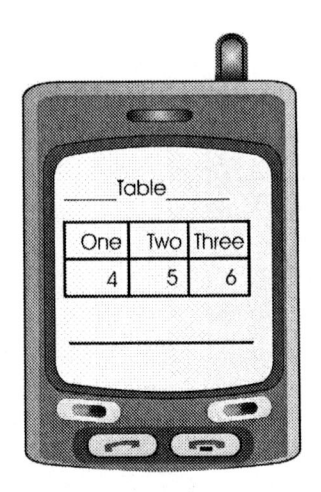

Figure 3.15 Table Example2

Prev

The <prev/> command causes a transfer or return to the previous card in the browser's history stack. If there is no previous card, then this command has no effect. The <prev/> command has no attributes. The <prev>...</prev> form may be used, but only when the command contains information such as set values for variables. When no information is in the command, the short form <prev/> should be used. Example 3.20 shows the use of the <prev/> command, and Figure 3.16 shows the screen for this example.

Wireless Markup Language (WML)

```
<?xml version="1.0"?>
<!DOCTYPE wml PUBLIC "-//WAPFORUM//DTD WML 1.1//EN"
"http://www.wapforum.org/DTD/wml_1.1.wml">
<wml>
<card id="card1" title="Prev Example">
<p>
This goes back to previous card.
<do type="prev" label="Back">
<prev/>
</do>

</p>
</card>
</wml>
```

Example 3.20

Figure 3.16 Prev Example

Chapter 3

Noop

The <noop/> command causes nothing to be done. It is a no operation, and the browser does nothing when it encounters this command. It is useful when a task is defined in the <template>...</template> command (discussed in Chapter 4), which causes the same command to be done in all cards in a deck, and if there is one card that you do not want the template command(s) to be done, you can insert the noop command in that particular card. The use of <noop/> will be discussed more in Chapter 4 when discussing the <template> command.

Summary

- WML is composed of cards and decks. Cards are individual screens and one or more cards compose a deck.

- A deck is identified by a URL address.

- WML documents delimit information with start and end tags. <wml>...</wml> are the start and end tags for a deck.

- The <card> command surrounds the contents of each card.

- Comments are surrounded by <!— comment —>.

- Variables have the form $var1 or $(var1). A variables name must start with a character or an underscore. Not a number or other special character.

- Text formatting is done using bold, italics, underlining, font size, and emphasizing. The formatting commands are: <b>, <i>,<u>,<big>,<small>, <em>, and <strong>.

- The <br/> command causes a line break to occur in the display.

77

Wireless Markup Language (WML)

- The <p> command defines the paragraph in the WML display.

- Commands have the form of <command>...</command> and attributes exist for each command. Attributes are listed within the brackets. <command attribute1="value1">.

- The <card>...</card> command defines the card in a WML deck.

- The <do>..</do> command starts cards tasks.

- The <go> command transfers to the specified URL.

- The <input> command is used to create a text field for the user to enter text.

- The <anchor> command is used to define a hyperlink. It can have the form <anchor>...</anchor> or <a>...</a>. The <a> form cannot be used to pass variables.

- The <table> command creates a table in a WML document. The rows are defined by <tr>...</tr> and the cells are defined by <td>...</td>.

- The <prev/> command causes a transfer back in the browser's history stack to the previous card.

- The <noop/> command causes no action to be taken by the browser.

Questions

1.) What is the basic structure of WML programs?

2.) What command delineates the WML program?

3.) How are comments inserted into the code?

4.) Is $3text a legal variable?

5.) What is valid for WML variables to start with?

6.) What is the <input> command used for?

7.) How would you get a sentence to be both bold and underlined?

8.) What command causes a line break to occur in the display?

9.) Write a valid <card> command with an id of "card1" and a title of "First Card".

10.) What attribute of the <card> command can be used to specify a URL that the browser will go to when this card is entered going backward.

11.) List five of the valid values for the type attribute of the <do> command.

12.) What is the required attribute for the <go> command?

13.) What value would the format attribute of the <input> command have in order to restrict the input to three capital letters followed by two numbers?

14.) When would you use the <anchor> command instead of the <a> command?

15.) What is missing in the following command: <table align="RRR" title="Table"> ?

16.) What command returns control to the previous URL in the browser stack?

Wireless Markup Language (WML)

Problems

1.) Write a program to say "Hello" on one line and "How are you?" on the second line.

2.) What is wrong with the program listed below?

```
<?xml version="1.0"?>
<!DOCTYPE wml PUBLIC "-//WAPFORUM//DTD WML 1.1//EN"
"http://www.wapforum.org/DTD/wml_1.1.xml">

<wml>
<!-- This example is
<!—for showing the use of the line break-- >
        >

<card id="card3" title="Line Break">

<p>

First new line <br/>

Second new line <br/><br/>

Third line
</p>
</card>

</wml>
```

3.) Write a program that has a deck with three cards, and when you enter card3 going forward, you return to card1.

4.) Write a program that has three cards in a deck, and each card has a forward button and a back button. (Use the <prev/> command when programming the back button.)

5.) Write a program that has a button labeled "Refer" and when it is pushed, you go to a different program on the same server named "wmlex44.wml".

6.) Write a program that has the user input his first name, and restrict it to capital letters with a "-" on each side of the name.

7.) What is wrong with the program below?

```
<?xml version="1.0"?>
<!DOCTYPE wml PUBLIC "-//WAPFORUM//DTD WML 1.1//EN"
"http://www.wapforum.org/DTD/wml_1.1.xml">

<wml>

<card id="cardone" title="Error Problem">

<p>
Hello!<br/>
Your phone number is:
<input type="accept" name="number" value="333-4444"/>
</p>

</card>

</wml>
```

Wireless Markup Language (WML)

8.) Write a program that has a link from the word "Next" on the first card that goes to the next card in the deck. On the second card have a link from the word "Initialize" to the program "ex55init.wml", and have it pass the variable "loop" set to a value of "1".

9.) Write a program with a table of three columns and two rows with the words "one","two","three","four","five","six" in each cell.

Chapter 4

WML Objects and Syntax

WML has more capabilities, commands, and features that are presented in this chapter. Appendix I lists all the commands of WML as well as each command's attributes.

Templates

The <template>...</template> command is used to place the same code in all of the cards in a deck. The code that is within the <template> tags will be executed on every card in the deck. If you want to override the template commands on one of the cards, you can put the same command (e.g. do or onevent) and use the same name attribute for that command, and then the commands in the card will be executed, not the commands in the template. This is where the <noop/> command is often used. If you have a command in the template that you do not want executed on one of the cards, by putting the command with the same name on the card and then using <noop/>, the <noop/> will override the template command. The <template> command and its contents are physically put within the <wml> tags but not in any specific card.

The **onenterforward**, **onenterbackward**, and **ontimer** attributes of the <template> command are used in the same fashion as these attributes when on the <card> command. When entering a card going forward (**onenterforward**) or backward (**onenterbackward**) control is transferred to the spec-

Wireless Markup Language (WML)

ified URL, or address or card number. **On timer** specifies what address or URL to go to when the timer times out. When used with a template command, these transfers will occur on all cards in the deck unless they are specifically overridden on certain cards.

Example 4.1 shows the use of the template command to put a "Previous" button on all cards, that executes the <prev/> command when pushed. However, on card1 a <do> command was inserted with a <noop/> command so that the first card would not have a "Previous" button. Also in this example, a <do> command with a "Forward" button was put on cards 1 and 2. Figure 4.1 shows the screens of the three cards as they occur from example 4.1.

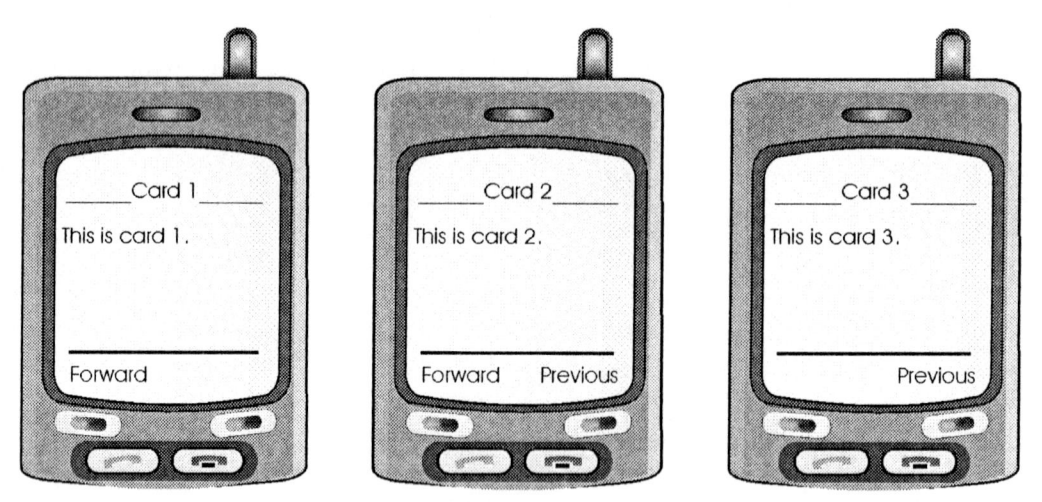

Figure 4.1, Template Example

Chapter 4

```
<?xml version="1.0"?>
<!DOCTYPE wml PUBLIC "-//WAPFORUM//DTD WML 1.1//EN"
"http://www.wapforum.org/DTD/wml_1.1.xml">

<wml>
<template>
 <do type="prev" label="Previous">
 <prev/>
 </do>

</template>

 <card id="card1">
<do type="accept" label="Forward">
 <go href="#card2"/>
 </do>
<do type="prev" label="Previous">
 <noop/>
 </do>
<p>
This is card 1.
</p>
</card>
 <card id="card2">
<do type="accept" label="Forward">
 <go href="#card3"/>
 </do>
<p>
This is card 2.
</p>
</card>

<card id="card3">
<p>
This is card 3.
</p>
</card>
</wml>
```

Example 4.1

Images

Images in a WML document are defined using the <img/> command. It has required attributes of **alt** and **src**. The **alt** attribute defines the text that is displayed by the browser if it cannot display the image. The **src** attribute gives the relative address or absolute URL address of the image. If the **localsrc** attribute is used, its value will replace the **src**'s value. The **localsrc** attribute specifies an internal representation for the image. If used, it specifies images stored on the WAP device as opposed to specifying images on the server. The src attribute specifies images stored on the server or at some other URL.

The **vspace** attribute is used to define empty vertical space around the image, and the **hspace** attribute defines the corresponding horizontal spaces. These attributes are given in pixels or percentages. The **align** attribute is the same as in HTML, and has allowable values of **top**, **middle**, and **bottom**. They define the position of the image relative to the text.

The size of the image is defined using the **width** and **height** attributes, in pixels or percentages. If **width** and **height** are not specified, the image comes in at its original size.

Example 4.2 shows the image command being used with the attributes of **src**, **alt**, **vspace**, and **hspace**, and Figure 4.2 shows the associated screens.

Chapter 4

```
<?xml version="1.0"?>
<!DOCTYPE wml PUBLIC "-//WAPFORUM//DTD WML 1.1//EN"
"http://www.wapforum.org/DTD/wml_1.1.xml">



<wml>

 <card id="card1" title="Image Example">
<p>

<img src="sunny.wbmp" alt="sunny" vspace="2" hspace="4"/>
</p>
</card>

</wml>
```

Example 4.2

Figure 4.2, Image Example

Wireless Markup Language (WML)

Currently the WAP standard has only one defined image format which browsers need to recognize: Wireless Bitmap (WBMP). Some WAP browsers also recognize GIF and JPEG formats, but usually mobile phones only accept WBMP. There are many programs available that convert images of other formats to WBMP. The Nokia Toolkit that is available on-line, converts JPEG and GIF images to WBMP. The whole emphasis of WBMP is small images and small number of pixels due to the small screens in mobile phones.

Passing Variables

WML allows the defining of variables which can then be used between cards within a deck, or between decks. This is not possible in HTML. The setting of variables and then passing them between cards can increase the flexibility of the functions done within decks or between decks. The variables are set with the <setvar/> command and the required attributes, **name** and **value**. Usually the <setvar/> command is used between a <go>...</go>, <prev>...</prev>, or a <refresh>...</refresh> command.

The **name** attribute of the <setvar/> command defines the identifier or name of the variable that WML code can refer to, using $variable or the $(variable) form. The **value** attribute sets the value of the variable.

Another common way to define and set the value of variables is with the <input/> command. The <input/> command allows for the user to input text and when the **name** attribute is used, that identifier is assigned the value of the inputted data, Also, in the <input/> command, the **value** attribute can be used to assign an initial value to the variable named in the same command.

Example 4.3 shows the passing of a variable between cards in the same deck. When card 1 is entered going forward, the variable State is set to the value "Initial". When the button "Change" is pushed, the variable State is set to "Changed". And, when the Return button is pushed, the variable State

```
<?xml version="1.0"?>
<!DOCTYPE wml PUBLIC "-//WAPFORUM//DTD WML 1.1//EN"
"http://www.wapforum.org/DTD/wml_1.1.xml">

<wml>

 <card id="card1" title="Var Display">
<onevent type="onenterforward">
<refresh>
 <setvar name="State" value="Initial"/>
</refresh>
</onevent>

<onevent type="onenterbackward">
<refresh>
 <setvar name="State" value="Returned"/>
</refresh>
</onevent>

<do type="accept" label="Change">
 <go href="#card2">
 <setvar name="State" value="Changed"/>
 </go>
</do>
 <p>
Card 1 current state: $State
</p>
</card>

<card id="card2" title="Var Display2">
<p>
 Card 2 current state: $(State)
</p>

<do type="accept" label="Return">
 <prev/>
</do>
</card>

</wml>
```

Example 4.3

Wireless Markup Language (WML)

is set to "Returned". Figure 4.3 shows the screens for Example 4.3. (Note that in Example 4.3 both forms of the variable "State" are used :$State and $(State). Technically the parentheses are only needed if necessary to separate the variable name from surrounding text.)

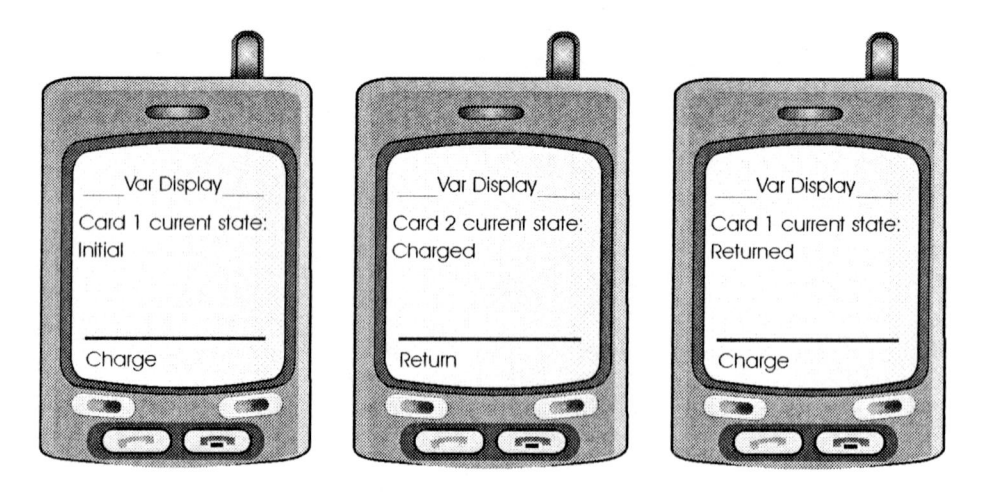

Figure 4.3, Variable Example

Probably the most common passing of variables is done between a WML deck and a WMLScript program. WMLScript is discussed in Chapter 5 and 6, but the passing of variables to WMLScript functions from WML decks is done by using the <setvar/> command or the <input/> command. In the WMLScript function, the same variable names are used and thus the passing of variables is very easy. As will be discussed in the chapters on WMLScript, the variables (or new variables defined in the WMLScript function) can then be easily passed back to the WML deck. Example 4.4 shows the code for passing a variable named "number" to a WMLScript function named "factorial" in a WMLScript program named "calculate.wmls". The variable is assigned in an <input> command, and the value is determined by the user input. Note that the variable "result" is used by the WMLScript function to return the value of the factorial calculation.

```
<?xml version="1.0"?>
<!DOCTYPE wml PUBLIC "-//WAPFORUM//DTD WML 1.1//EN"
"http://www.wapforum.org/DTD/wml_1.1.xml">

<wml>

 <card id="card1" title="Factorial">
<do type="accept" label="Calc Fact.">
<go href="calculate.wmls#factorial($number)"/>
</do>

<p>
Number:
<input name="number" title="Number:"/>
<br/>
=<u>$result</u>
</p>

</card>

</wml>
```

Example 4.4

Wireless Markup Language (WML)

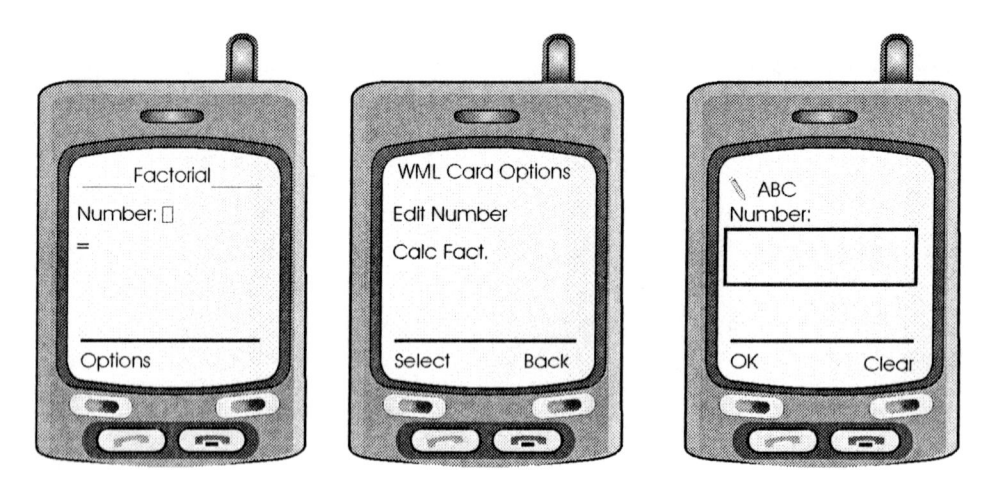

Figure 4.4, Variable Example 2

More Commands

Select

The <select>...</select> command is used to create selection lists. Options in the list are defined with the <option>...</option> command, or the <optgroup>...</optgroup> command. Each <select> tag has one or more <option> tags, or <optgroup> tags within it that contain choices that the user must select from. The browser displays the text associated with each option. The **multiple** attribute in the <select/> command is used to specify if more than one selection is allowed. The default value is **false**, but if **multiple** is set to **true**, then multiple choices from the list can be made.

The **name** attribute of the <select/> command is required and defines the name or identifier for the variable selected in the list. The **value** attribute is used to specify the default selection. When the user makes a choice, the **value** attribute of the <option/> command becomes the value assigned to the variable. In the case when the **multiple** attribute is set to **true**, the different variable values are separated by semicolons.

The **iname** attribute of the <select/> command defines an alternative name for the variable that receives the number of the selection made. So, if there are three options listed, and the third option is selected, the variable name specified by **iname** is set to 3. The **ivalue** attribute is the initial value used for the variable specified in the **iname** attribute. If both **name** and **iname** and **value** and **ivalue** are specified, the **iname** and **ivalue** are used instead of the **name** and **value** attributes.

The **title** attribute of the <select> command specifies the title that appears as the title of the selection list. The **tabindex** attribute specifies the order of the list elements with non-negative integers. This attribute is currently not supported in some of the browsers, such as Phone.com's browser.

Example 4.5 shows the use of the <select> command with the **name**, **iname**, **value** and **ivalue** attributes. Since **ivalue** takes precedence over **value**, the screen in Figure 4.5 shows that the initial value of the variable is set to the value specified in **ivalue**, not **value**. On the first screen, the mode is set to "car" not "plane". The second screen shows what is seen after pressing the "Select" button. And the third screen shows the display after choosing the second choice on the list and then pushing "OK".

Wireless Markup Language (WML)

```
<?xml version="1.0"?>
<!DOCTYPE wml PUBLIC "-//WAPFORUM//DTD WML 1.1//EN"
"http://www.wapforum.org/DTD/wml_1.1.xml">

<wml>
 <card id="card1" title="Value Selection">
<p>
Pick Trans.: <br/>
<select name="mode" iname="imode" value="plane" ivalue="3">
<option value="plane">Airplane</option>
<option value="train">Railway</option>
<option value="car">Automobile</option>
</select>
<br/>
Mode: $(mode)
Index: $(imode)

</p>
</card>

</wml>
```

Example 4.5

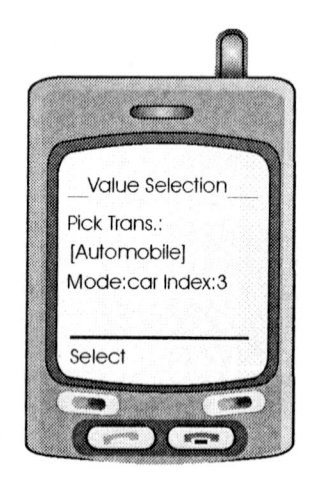

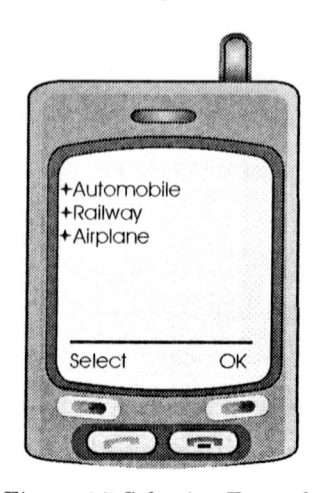

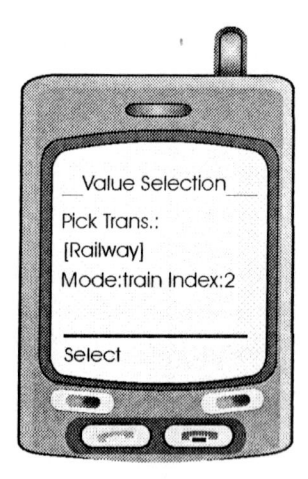

Figure 4.5, Selection Example

Option

The <option>...</option> command is used to define options in a selection list. The value of the **value** attribute is assigned to the variable named in the <select> command, when the user makes the selection associated with that value.

The **title** attribute of the <option> command specifies the title in the browser for the selected option. The browser will determine the way the title text for the selected options is displayed on the screen. Some browsers, such as the Nokia WAP Toolkit and the Ericsson MC 218 will not display the titles of the selected options on the screen if browser-displayable text has been specified within the <option>...</option> command. Example 4.6 shows the use of the <option> command and the <select> command.

Figure 4.6 shows the screens associated with example 4.6, and you will note that the title attribute is not displayed, but instead the text between the <option> tags is displayed.

```
<?xml version="1.0"?>
<!DOCTYPE wml PUBLIC "-//WAPFORUM//DTD WML 1.1//EN"
"http://www.wapforum.org/DTD/wml_1.1.xml">

<wml>
 <card id="card1" title="Option Example">
<p>
<select name="Tools">
<option value="axe">Axe</option>
<option value="saw">Saw</option>
<option value="hammer">Hammer</option>
</select>

</p>
</card>

</wml>
```
Example 4.6

Wireless Markup Language (WML)

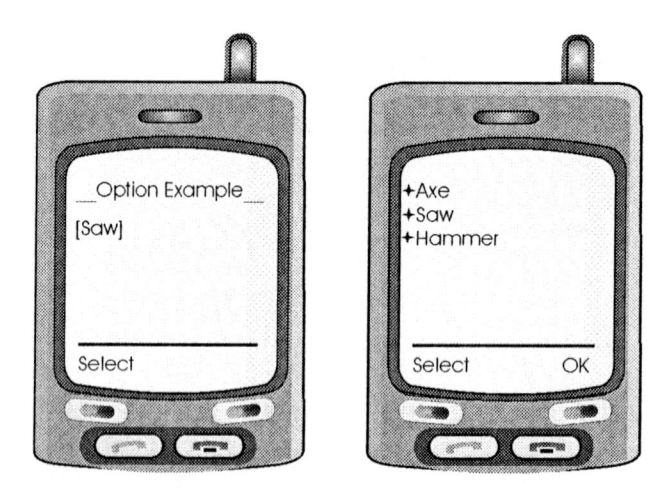

Figure 4.6 Option Example

The **onpick** attribute of the <option> command is used to specify a URL to which the browser will transfer if this option is selected.

Optgroup

The <optgroup>...</optgroup> command is used to group the <option> elements into a hierarchical list. The user will see the <optgroup> selections first, and then go to another level for the <option> selections associated with each <optgroup> element. The **title** attribute of the <optgroup> command is used to specify a title for the selection group to be displayed in the browser. Example 4.7 shows the use of the <optgroup> command. Figure 4.7 shows the screens for this example. The second screen shows the <optgroup> groups that each has their own options list. The third screen shows the list for the Aerobic Tests group. (The screen is not shown for the Strength Tests.)

```
<?xml version="1.0"?>
<!DOCTYPE wml PUBLIC "-//WAPFORUM//DTD WML 1.1//EN"
"http://www.wapforum.org/DTD/wml_1.1.xml">

<wml>
 <card id="card1" title="Optgroup Example">
<p>

Pick a test:<br/>

<select name="Test">

<optgroup title="Aerobic Tests">

<option value="run">Run Test</option>
<option value="walk">Walk Test</option>
<option value="bike">Bike Test</option>
</optgroup>

<optgroup title="Strength Tests">

<option value="situp">Situp Test</option>
<option value="pushup">Pushup Test</option>
<option value="pullup">Pullup Test</option>
</optgroup>
</select>
</p>

</card>

</wml>
```

Example 4.7

Wireless Markup Language (WML)

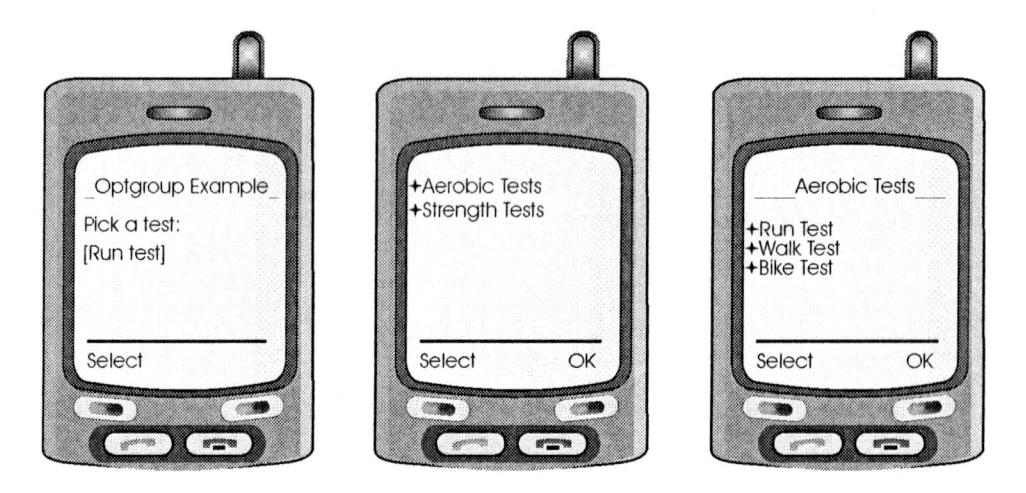

Figure 4.7 Optgroup Example

Refresh

The <refresh>...</refresh> command is used to refresh an active document. The command is always used within an event so the tasks or the <setvar> command within the <refresh>...</refresh> commands are carried out whenever the event occurs. This means that the browser carries out a task or assigns an initial values to variables when the task is performed. The <refresh> command has no attributes.

<refresh> command is very useful when initializing variables for WMLScript functions.

Example 4.8 shows the use of the <refresh> command, where the variables "number" and "result" variables are initialized to "0.0" every time this card is entered going forward. (Due to the onenterforward event). Figure 4.8 shows the screens for this example.

Chapter 4

```xml
<?xml version="1.0"?>
<!DOCTYPE wml PUBLIC "-/WAPFORUM//DTD WML 1.1//EN"
"http://www.wapforum.org/DTD/wml_1.1.xml">

<wml>

 <card id="card1" title="Refresh Example">
<onevent type="onenterforward">
<refresh>
<setvar name="number" value="0.0"/>
<setvar name="result" value="0.0"/>
</refresh>
</onevent>
<p>
Number is <b> $(number)</b><br/>
Result is <b> $(result)</b><br/>
</p>
</card>

</wml>
```

Example 4.8

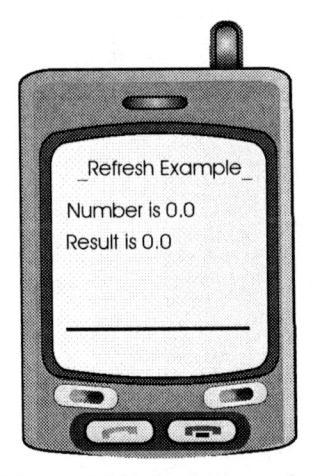

Figure 4.8 Refresh Example

Timer

The <timer/> command causes an application to initiate a given action after a certain period of time. The timer is initialized and started when opening a card, and the timer stops when the card is closed. The timer will begin counting down until it times out, or the card is left. The value of the timer needs to be a positive integer, and each count represents one-tenth of a second. If the user is still on the card when the timer times out, a pre-determined action takes place. And, you can have only one timer on each card.

A timer is often used to display an advertisement or a logo page for a few seconds, before you automatically navigate the user to the start page of the application.

The **value** attribute of the <timer/> command is a required value. It specifies the timing period in tenths of seconds. It needs to be a positive integer. The **name** attribute is used to link a variable to the timer, and the variable is used to store the value of the timer. This **name** attribute can be used if it is necessary to pass the timer as a variable between documents.

In example 4.9 the logo ":ABC Inc" is displayed for 10 seconds before going to the first card in the main application. Figure 4.9 shows the screens for this example.

Figure 4.9 Timer Example

Chapter 4

```
<?xml version="1.0"?>
<!DOCTYPE wml PUBLIC "-//WAPFORUM//DTD WML 1.1//EN"
"http://www.wapforum.org/DTD/wml_1.1.xml">

<wml>

 <card id="cardlogo" title="Timer Example" ontimer="#card1">
<timer value="100"/>

<p align="center">
<b><u><big>ABC </big></u></b><br/>
<b><u>Inc.</u></b>
</p>
</card>

</wml>
```

Example 4.9

Fieldset

The <fieldset>...</fieldset> command is used for grouping elements in a display. It can group <input> commands, <select> commands, or just text. The **title** attribute of the <fieldset> command specifies the title for the group that shows on the display. However, it should be noted that not all browsers group elements based on the <fieldset> command. Some ignore it.

Example 4.10 shows the use of the <fieldset > command to group the requested inputs into two groups, "Personal Info", and "Health Info".

The Nokia WAP Toolkit does not display the <fieldset> command with its titles, and that is why Figure 4.10 shows how example 4.10 would look on a PDA, or a device whose browser recognizes the fieldset command.

101

Wireless Markup Language (WML)

```
<?xml version="1.0"?>
<!DOCTYPE wml PUBLIC "-//WAPFORUM//DTD WML 1.1//EN"
"http://www.wapforum.org/DTD/wml_1.1.xml">

<wml>

 <card id="card1" title="Fieldset Example">
<do type="accept" label="Save">
<go method="post" href="http://wap.acta.fi/cgi-bin/wap/medical.pl">
<postfield name="age" value="$(age)"/>
<postfield name="sex" value="$(sex)"/>
<postfield name="pulse" value="$(pulse)"/>
<postfield name="temp" value="$(temp)"/>

</go>
</do>
<p>
<fieldset title="Personal Info">
Age (years): <input type="text" name="age" format="*N"/>
Sex:
<select name="sex">
 <option value="Woman">Woman</option>
 <option value="Male">Man</option>
</select>
</fieldset>

<fieldset title="Health Info">
Pulse: <input type="text" name="pulse" format="*N"/>
Temp <input type="text" name="temp" format="*N"/>
</fieldset>

</p>
</card>

</wml>
```

Example 4.10

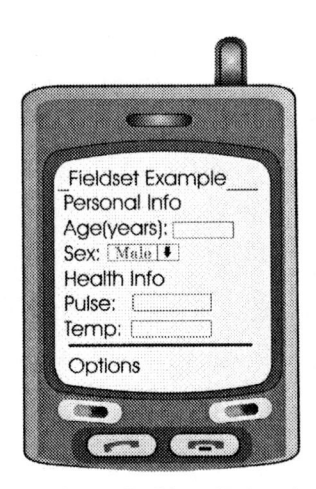

Figure 4.10 Fieldset Example

Head

The <head>...</head> command contains meta elements related to the WML deck. The <head> command can have either <access> or <meta> commands within it. The <head> command cannot be empty, and it must precede any <card> commands in the deck. Commands and text within the <head>...</head> command are not displayed by the browser. The <head> command has no attributes.

Meta

The <meta/> command contains general meta elements related to the WML deck. Just like in WWW sites, search engines can use this information defined with the <meta/> command that is invisible to the browser. The <meta/> command has no matching end tag, and is only used in the <meta/> form. The attributes of the <meta/> command are **content, name, http-equiv, forua,** and **scheme.**. The **content** attribute is used to define the value of the property containing the meta element, which is linked to either the **name, http-equiv** or **forua** attribute. The **name** attribute specifies a name for the property — the browser will ignore this attribute. The **http-equiv** attribute can be used instead of the **name** attribute to show the browser that a property is to be treated in the same way as header information is treated in HTTP. The **forua** attribute defines a property as visible (true) or invisible (false) to the browser. The **scheme** attribute is used to define format or composition as name-value pairs, which can be used when interpreting property data.

In example 4.11, the maximum age of a deck is defined by giving the **http-equiv** attribute the value "Cache-Control", and the **content** attribute the value "max-age=3600", which refers to the time (in seconds) that the document stays in the cache memory. The default time is 30 days, but this can be redefined using this command. After this time, the browser does not look in the cache memory for the document, but goes to the server to retrieve it. If the max-age value is set to zero, the browser will always retrieve the document from the server.

```
<head>
 <meta content="max-age="3600" http-equiv="Cache-Control"/>
</head>
```

Example 4.11

There is no associated figure for this example since none of this is displayed by the browser.

Access

The <access/> command defines the visibility of the whole deck. If the deck does not contain an <access/> command, the user can move to this WML deck from cards in any deck. A deck may contain only one <access/> command.

The attributes of the <access/> command are **domain, path, id** and **class**. The **domain** and **path** attributes are used to define decks from where the deck can be accessed. The browser always controls this information when the user moves from one deck to another. The browser starts checking the authorized URL address from the end of the **domain** attribute and the beginning of the **path** attribute. In practice this means that *wap.acta.fi* matches the domain address *acta.fi* but not the address *ta.fi*. Also, */web/wap* matches the **path** attribute value */web*, but not the **path** attribute value */webpages*. The **path** attribute also accepts a relative address as its value, in which case the browser converts it to an absolute path before the comparison.

In example 4.12, visibility is allowed in all locations of the WAP directory (and its subdirectories) in the *wap.acta.fi* domain. This means that addresses like *acta.fi/WAP/test.wml* and *wap.acta.fi/WAP/WEB/public/test3.wml* are allowed. The default value of the **domain** attribute in the <access/> command is the domain address of its present location, and the default value of the **path** attribute is "/", which is the root path of the present deck.

```
<head>
<access domain="wap.acta.fi" path="/WAP" />
</head>
```

Example 4.12

There is no associated figure since none of the <head> contents are displayed on the screen.

Postfield

The <postfield/> command is used to define name-value pairs that are used to send information from the browser to the server. The **name** attribute is used to define the field name that will contain the value of the **value** attribute, which in most cases is the value of a form field variable. This command is used in combination with WML form elements (e.g. <input>) in CGI and servlet programming. This is how variables and their values are most often inputted in forms in the browser and then sent to the server. Example 4.10 shows the use of the <postfield/> command to pass name-value pairs from the browser to the server-based perl program, "medical.pl".

Onevent

The <onevent>...</onevent> command is used to have a command or commands executed upon the occurrence of a specific event, as defined by the **type** attribute of the <onevent> command. The **type** attribute can be onenterforward, onenterbackward, or ontimer. If onenterforward is used, for example, whatever commands are within the <onevent>...</onevent> tags will be executed when the card is entered going forward. Typical commands used within the <onevent>...</onevent> tags are <setvar>, <refresh>, or <go>.

Often, the same action can be done by using the **onenterforward, onenterbackward**, or **ontimer** attributes of the <card> command, but if you have to set variables, the <onevent> command is used since it allows for several commands to be used within the <onevent> tags.

Example 4.8 shows the use of the <onevent> command with the **type** attribute set to onenterforward. In this example, the variables "number" and "result" are set to the value "0.0" when the card is entered going forward, using the <refresh> command.

Example 4.13 shows the use of the <onevent> command with the <go> command to go to the WML program "wmlex6.wml", when the card is entered going backwards. Figure 4.11 shows the screen for example 4.13, except the <onevent> doesn't affect the display of the card shown.

Chapter 4

Figure 4.11 Onevent Example

Summary

- The <template> command places the same code in all of the cards in a deck.

- Images are used with the <img> command, and the WAP standard image format is the WBMP. However, some WAP browsers also recognize gif and jpeg formats as well.

- Variables can be used between cards within a deck, between decks, or between cards and WMLScript or cgi programs. The <setvar> command, and the <input> commands are the most common way of setting variables.

- The <select> command is used to create selection lists. The options in the list are defined with the <option> command or the <optgroup> command.

- The <option> command defines the options in a select list.

Wireless Markup Language (WML)

- The <optgroup> command lets you make a hierarchical selection list, by grouping options.

- The <refresh> command refreshes an active document. This command is always used within an event.

- The <timer> command initiates an action after a certain period of time. The timer is initialized and started when a card is opened.

- The <fieldset> command groups elements in a display.

- The <head> command contains meta elements related to the WML deck, and are similar to the HTML meta elements.

- The <meta> command contains general meta elements for the WML deck.

- The <access> command defines the visibility of the whole deck.

- The <postfield> command defines name-value pairs that are used to send information from the browser to the server.

- The <onevent> command defines certain commands that are executed when a specific event occurs, such as, onenterforward, onenterbackward, or ontimer.

Chapter 4

Questions

1.) How do you override the <template> command on a specific card?

2.) When using the localsrc and src attributes on an image, which one takes precedence?

3.) Which image format is a WAP standard?

4.) What attribute in the <setvar> command defines the identifier of the variable?

5.) Where does the most common passing of variables occur?

6.) How do you create selection lists in WML?

7.) What attribute is used in order to allow more than one selection to be chosen?

8.) How do the <option> command and the <optgroup> command differ?

9.) When would you use or need the <refresh> command?

10.) What does the value of the timer represent?

11.) What is the <fieldset> command used for?

12.) Within what command tags does the <meta> command occur?

13.) How would you restrict the ability of someone to move to a specific deck?

14.) When is name-value pairs most often used?

15.) When would you use the <onevent> command as opposed to the onenter forward, etc attributes of the <card> command?

Wireless Markup Language (WML)

Problems

1.) Write a deck that uses the <template> command to put a "previous" command on each of 3 cards in the deck. Also, put a "forward" button on each card, and no "previous" on the first card.

2.) Write a program that puts an image of a rain cloud ("rainy.wbmp" in the nokia WAP Toolkit) in the center of a page, and use the alt attribute to enter the name "rainy" if not displayed.

3.) Write a program that sets variables "height" and "width" in card one, using the <refresh> command, and then displays these variables on card two.

4.) Write a program that has the user enter the values for the variables "name" and "phone" on the first card in a deck, and then displays them on the second card in the deck.

5.) Write a program that sets up a selection list for three different kinds of semiconductors : BJT, FET, and CMOS, and then each choice go to a different card in the deck that is titled BJT, FET, or CMOS.

6.) Using the <optgroup> command, set up a program that shows two main choices of vehicles, "cars" or "trucks", and then under each of these selection show a list of 3 cars by make and 3 trucks by make.

7.) Write a program that uses a timer to display the first card in a deck for 10 seconds which shows a company name in large and bold print, and then automatically transfers to a second card with the Company name, address and phone number.

8.) Use the <fieldset> command to form two groups of inputs on the display, one for color choice (red, blue or white) and the other for size (small, medium, or large).

9.) Write a program that when the second card is entered going forward, it goes to the fourth card. And, when the third card is entered going backward, it goes to the first card. Put in all appropriate forward and back buttons on all cards.

Chapter 5

WMLScript Programming

WMLScript is a simple scripting language similar to JavaScript. The biggest difference between the two is that WML contains references to the URL address of a WMLScript function, whereas JavaScript functions are usually embedded in the HTML code.

WMLScript is a object-oriented, procedural scripting language whose purpose is to enhance the abilities of WML. WMLScript is part of the application layer of the WAP architecture and can be used to add procedural logic to WAP devices, either as part of WML documents or as a standalone tool in independent applications.

Scripting Structure

WMLScript is a language with a weak type definition, which means that variable types are not defined during declaration; the data types are determined during execution. The basic information types of WMLScript are: Boolean, integer, floating point, string and invalid. WMLScript automatically converts values between data types if necessary.

WMLScript is based on ECMAScript (also known as ECMA-262), which was conceived by the European Computer Manufacturers Association in 1998, and is based on Javascript 1.1.

Wireless Markup Language (WML)

WMLScript includes a number of operators (assignment operators, arithmetic operators, etc), which are fairly similar to those in JavaScript. WMLScript also supports locally installed standard libraries, which are currently **Lang**, **Float**, **String**, **URL**, **WMLBrowser**, and **Dialogs**.

```
// This is a comment in a single row.

/*
 This is a multiple row comment
 and nested comments are not
 allowed.
*/
```

Example 5.1

The comments in WMLScript are similar to those in JavaScript. A single-row comment begins with // and multiple-row comments begin with /* and ends with */. Nested comments are not allowed. Example 5.1 shows some valid comments.

Functions

WMLScript includes functions, which are called if a task calling the functions has been linked to WML deck events, and the event (such as pushing a button) occurs. Parameters can be passed to the functions, and other functions (within the same code, in another compilation unit or library) can be called from within the functions. After a WMLScript has been executed, the system usually returns to the location from which the call was made. So, after calling a WMLScript function from a WML deck, the system returns to the WML card that called the function.

A WMLScript function is declared in the compilation unit by using the function name and a parameter list. The function name can be preceded by the reserved word **extern**, which makes the function externally accessible, as from a WML document where the function is called. The parameter list can be empty or it may contain any number of parameters passed for the function. After the function declaration, the program code follows, enclosed in the {} brackets. When a function is called, the code enclosed in the {} brackets is executed.

Functions can call other functions, whether inside the same compilation unit, in other compilation units or in libraries. WMLScript functions cannot be nested. Function names within a compilation unit must be unique, but different units may contain functions with the same name. This is because the WMLScript interpreter places the full URL address in front of the function name, with the compilation's unit name at the end of the URL.

When calling a function – usually from a WML document – the parameter list of the calling document has to match the function declaration with the exact same number of parameters, and in the same order. The parameters are separated by commas. The number of parameters is not limited, and a function may have no parameters, as well. The function still is followed by ( ) even if no parameters are passed.

The function always returns a value, even when this is not specified with a return statement. A return statement makes the function return the required value to the calling document. If no return statement is used, then the function by default returns an empty string.

Example 5.2 shows a WML deck with one card that gets a number inputted from the user, and then passes the variable "number" to a WMLScript function that calculates the factorial of the number passed, and then returns the answer in the variable "result".

Wireless Markup Language (WML)

```
<?xml version="1.0"?>
<!DOCTYPE wml PUBLIC "-//WAPFORUM//DTD WML 1.1//EN"
"http://www.wapforum.org/DTD/wml_1.1.xml">

<wml>

 <card id="card1" title="Factorial">

<do type="accept" label="Calc. Fact.">
<go href="wmlex51.wmls#factorial($(number))"/>
</do>
<p>
Number:
<input format="*N" name="number" title="Number"/>
<br/>
The result is:
<u>$(result)</u>
</p>

</card>

</wml>
```

Example 5.2 WMLScript Code

In the WML code, the <go href> command specifies the name of the WMLScript program, "ex51.wmls", followed by the name of the function – factorial - preceded by a # sign. One WMLScript program may obviously have multiple functions. Also, the same <go href> command passes the variable "number" as the input to the function named "factorial". Later in the WML code the variable "result" is displayed. This variable "result" is defined in the WMLScript code by the WMLBrowser.setVar command, and then passed back to the WML code. These library functions will be discussed later in the next chapter.

Chapter 5

Figure 5.1 displays the screens that one sees when executing the code in example 5.2

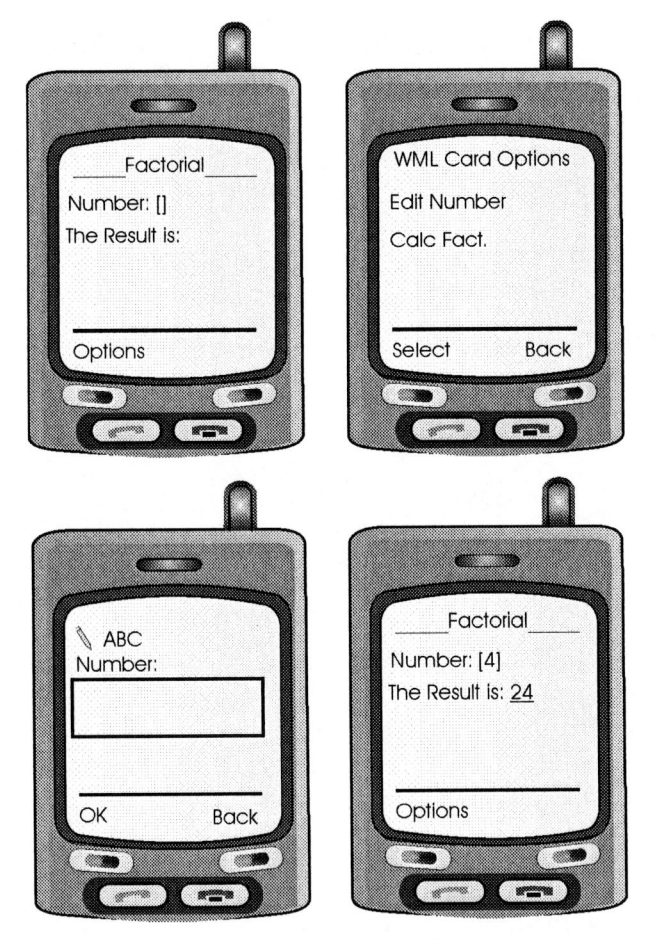

Figure 5.1 Factorial Example

WMLScript Variables

Four types of values can be declared for WMLScript variables: integer, floating point, string and Boolean. Because the data types are only supported internally, they are not defined during declaration, but the type is determined during the execution of the function. WMLScript converts between the types if necessary.

Wireless Markup Language (WML)

Integers can be presented in three different ways: decimal, octal, or hexadecimal values. Floating point values can be expressed with a decimal point or an exponent. Strings consist of 0-n characters, and are enclosed within inverted commas (" ") or apostrophes (' '). Some special marks can be expressed in ways that are similar to C and Java languages. An apostrophe inside a string can be marked as \', an inverted comma \", a backslash \\, a forward slash as \/, a line break as \n, and a tab as \t. A truth value in WMLScript can be either true or false. Example 5.3 shows some valid variables.

Variable identifiers are strings of unlimited length that start with a letter or an underscore, and continue with letters, digits, or underscores. Also, a variable identifier cannot be a reserved word, or a truth value – **true** or **false**.

Valid Variables

variable1 // this is a variable

Variable1 // this is a different variable

_variable2 //a valid variable

Const_Value // valid variable

a_b_c //valid variable

Example 5.3

Example 5.4 lists several invalid variables.

---

access
function
while
009
$variable
9variable

Example 5.4

Chapter 5

Reserved words in the WMLScript language are: **access, agent, break, continue, div, div=, domain, else, equiv, extern, for, function, header, http, if, isvalid, meta, name, path, return, typeof, url, use, user, var and while.** Reserved words which are not yet in use are: **delete, in, lib, new, null, this, void,** and **with.**

Variable declaration is compulsory in WMLScript. A variable is introduced with the command **var** , followed by the identifier or name, Variables must be declared before they can be used in the program code. The variable value can be initialized, but if this is not done, the value of the variable will be an empty string (""). The type of a variable is not specified in the declaration, but is determined during processing.

Example 5.5 shows a WML document that gets a number inputted by the user, and then calls the function "root" in wmlex55.wmls. The parameter "number" is passed in the calling of the "root" function in the <go href> command. And, the answer comes back in "result", which is defined in the WMLScript function using the WMLBrowser.setVar command. The function calculates the square root of the number inputted, using the sqrt function in the Float library. These libraries will be discussed in the next chapter.

117

```
<?xml version="1.0"?>
<!DOCTYPE wml PUBLIC "-//WAPFORUM//DTD WML 1.1//EN"
"http://www.wapforum.org/DTD/wml_1.1.xml">

<wml>

 <card id="card1" title="Calculation">

<do type="accept" label="Calc. Sq. Root">
<go href="wmlex55.wmls#root($(number))"/>
</do>
<p>
Number:
<input format="*N" name="number" title="Number"/>
<br/>
The Square Root is:
<u>$(result)</u>
</p>

</card>

</wml>


 *** WMLScript File ***

 extern function root(number) {
  var answer;
 if (number < 0) {
  answer = number;
 }
 else {
  answer = Float.sqrt(number);
 }

 WMLBrowser.setVar("result",answer);
 WMLBrowser.refresh();
```

Example 5.5

Chapter 5

Figure 5.2 shows the screens associated with example 5.5

Figure 5.2 Square Root Example

WMLScript supports two numeric variable types: integer and floating point. The size of the integer type variable is 32 bits. These values can be determined by using the **Lang** library functions **maxInt()** and **minInt()**. If a value is entered that is too great or too small, an error is generated and the variable is converted to the type **invalid**.

119

Wireless Markup Language (WML)

For floating point numbers, WMLScript supports 32 bit precision. The min and max values allowed can be determined by using the **maxFloat()** and **minFloat()** in the **Float** library. If a floating point value is entered that is too great or too small, an error is generated and the variable is converted to the type **invalid**.

WMLScript supports strings, which can contain letters, digits, and special characters.

It is also possible in WMLScript to define Boolean values for variables. In these cases the variable may have a the value **true** or **false**.

Another way to pass variables between WML code and WMLScript functions is to define the variable in the WML code, as in an input statement, but not pass any variable name in the <go href> command. Leave the variable field empty. Also, therefore the WMLScript function would contain no variable names in the first line after the function name. But, in the WMLScript code, the command WMLBrowser.getVar would be used to retrieve the variable. Example 5.6 is the same problem as shown in example 5.5, except the variable is passed using the WMLBrowser.getVar command.

The screens associated with example 5.6 are exactly the same as example 5.5, in figure 5.2.

```
<?xml version="1.0"?>
<!DOCTYPE wml PUBLIC "-//WAPFORUM//DTD WML 1.1//EN"
"http://www.wapforum.org/DTD/wml_1.1.xml">

<wml>

 <card id="card1" title="Calculation">

<do type="accept" label="Calc. Sq. Root">
<go href="wmlex56.wmls#root()"/>
</do>
<p>
Number:
<input format="*N" name="number" title="Number"/>
<br/>
The Square Root is:
<u>$(result)</u>
</p>
</card>
</wml>

*** WMLScript File ***

extern function root() {
 var number = WMLBrowser.getVar("number");
 var answer;
if (number < 0) {
 answer = number;
}
else {
 answer = Float.sqrt(number);
}

WMLBrowser.setVar("result",answer);
WMLBrowser.refresh();
}
```

Example 5.6

Wireless Markup Language (WML)

Operators

In WMLScript, the assignment operator is the equals sign (=). The assignment operator is used to assign a value to a variable. It assigns the value on the right side of the equals sign to the variable on the left side of the equal sign.

Example 5.7 shows the assignment first of an empty string to a variable string1, and then the assigning of the string "North" to a variable string2.

```
var string1 = "";

var string2 = "North";
```
Example 5.7

WMLScript also allows the use of short assignment operations, where a short calculation precedes the assignment. All short assignment operations can also be presented in the full-length format as well. Example 5.8 shows both the short assignment operation as well as the full-length assignment.

Arithmetic Assignment Operators in Short and Full-Length Formats

```
var x = 3;
x = x+5; // addition and assignment
x += 5;
x = x – 5; // subtraction and assignment
x -= 5;
x = x * 5; // multiplication and assignment
x *= 5;
x = x / 5; //division and assignment
x /= 5;
x = x div 5; // division with an integer and assignment
x div= 5;
x = x %5; // Modulo and assignment
x %= 5;
```
Example 5.8

WMLScript also contains bit operators such as listed in example 5.9. These operators shift and move bits.

Bit Shift Operations

<< Shift to left
\>> Shift to right
\>>> Shift to right (sign to positive value)
& And (returns true, if both operands are true)
| Or (returns true, if at least one of the operands is true)
^ Exclusive Or (returns true if one but not both are true)

Example 5.9

WMLScript language also contains unary value-addition operators, such as ++, which increases the value of the operand by 1. And, the operator – which decreases the value of the operand by 1. These value-addition operators may be used in the prefix or postfix notation. In the prefix notation, the addition or subtraction is done before the operand's value is evaluated. In the postfix notation, the operand is evaluated and then the addition or subtraction is done.

Example 5.10 lists the logical operators of the WMLScript language.

Logical Operators

! Negation (not). Returns true, if operand is false.

&& And (returns true, if both operands are true)

|| Or (returns true, if at least one of the operands is true)

Example 5.10

Wireless Markup Language (WML)

There is only one tertiary operator in WMLScript, and it is ?: . This is short form for the **if-else** expression. Example 5.11 shows the use of the normal if-else, which will be discussed in the next section, and the short tertiary operator form.

```
var a, b =6, c = 7;

if (b < c ) {
  a = c;
}

else {
  a = b:
}

// Short form with the conditional operator:

a = (b < c ? c : b) ;
```

Example 5.11

WMLScript supports the linking of strings using arithmetic sum operators (+). Linking of strings can also be done using the short assignment operator += , which works the same way as when combined with mathematical variables.

Comparison operators are used to compare two values. The operator != returns the value true when the two operands are not equal. The comparison operators for WMLScript are listed in example 5.12

Chapter 5

Comparison Operators

== equal

!= not equal

\> greater than

< less than

\>= greater than or equal

<= less than or equal

Example 5.12

Comparison of numerical values is straightforward. However, for Boolean type truth-values, the **true** value is larger than **false**. And, for strings the operators compare the alphabetical order of the strings; a string that comes first alphabetically, is treated as lower than a string that comes later.

The WMLScript language also has an operator called **isvalid**, which can be used to test whether a value (a literal or a return value of a statement) is valid (integer, floating-point value, string or a straight or convertible truth value) or not (its data type is isvalid). The operator returns either **true** or **false**, depending on whether or not the entry is valid (i.e. can be converted to a WMLScript internal data type). The operator does not reflect whether the entry is of the type the user intended. Example 5.13 shows the use of the **isvalid** operator.

Wireless Markup Language (WML)

```
var i = 5;

var str = "Hello!";

var a = isvalid i; // true

a = isvalid str; // true

a = isvalid (10/0) ; // false, can't divide by 0.
```
Example 5.13

The WMLScript language does not include arrays, but some functions in the **String** standard library treats library strings as character arrays. These are described in the next chapter.

Statements and Expressions

All commands in the WMLScript language end with a semi-colon. Program code inside the { and } characters is called a block. The components of a function are included in a program block, and other program code components — normally loop and selection statements — are also enclosed in blocks. After declaration, variables are visible until the end of the function in which they were declared. The scope of their visibility is not restricted to the block where the variable was declared.

WMLScript has an **if-else** structure. If the condition following the **if** statement is true, the block of code between the { and } brackets will be executed. If the condition is **false**, the browser will execute the statement following the } bracket of the if block. An optional **else** statement can be specified after the **if** statement, and will be executed in case the tested if condition is false. Example 5.14 shows a sample of an if-else statement.

Chapter 5

```
                    The if-else Structure

Function statement (number){
   Var result;

   If (number < 0) {
       Result = number;
       }
else {
   result = Float.sqrt (number);
}
return result;
}
```
Example 5.14

The if and else blocks may contain several executable statements, and the if structures can also be nested, as shown in example 5.15. Example 5.15 first shows the WML code to call the function "compare" which then shows the WMLScript with nested if statements, Figure 5.3 shows the associated screens for this code.

Wireless Markup Language (WML)

```
<?xml version="1.0"?>
<!DOCTYPE wml PUBLIC "-//WAPFORUM//DTD WML 1.1//EN"
"http://www.wapforum.org/DTD/wml_1.1.xml">

<wml>

 <card id="card1" title="Calculation">

<do type="accept" label="Compare">
<go href="wmlex515.wmls#compare($(number))"/>
</do>
<p>
Number:
<input format="*N" name="number" title="Number"/>
<br/>
The answer is:
<u>$(answer)</u>
</p>

</card>

</wml>
extern function compare (number) {
var mynumber = 10;
var result;
if (number < mynumber) {
 result = "Your number is smaller.";
}
else if (number == mynumber) {
result = "Our numbers are equal.";
}
else{
result =" Your number is bigger.";
}
 WMLBrowser.setVar("answer",result);
WMLBrowser.refresh();
```

Example 5.15

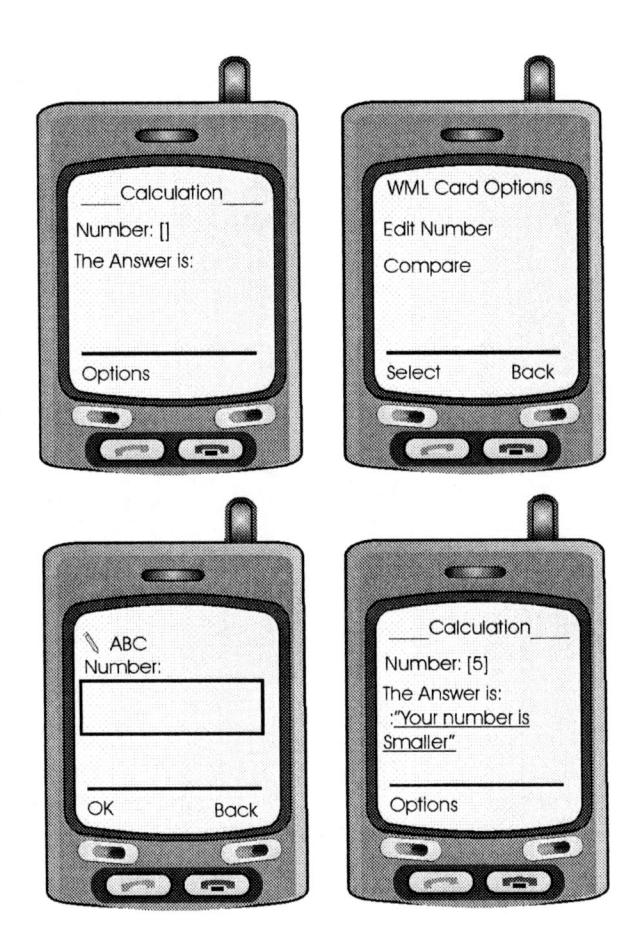

Figure 5.3 Nested If Example

WMLScript does not have a case-type statement, but chained if statements can be used for the same effect.

WMLScript includes a loop statement **while**, which tests for the loop condition before executing the loop block, and a stepwise loop statement **for.** The while statement is used when part of a program code is to be repeated for as long as a certain condition is true. If the condition is false when arriving at the statement, the while block is ignored, and the program execution continues with the code that follows that block.

Wireless Markup Language (WML)

Example 5.16 shows WML code calling a WMLScript function with a **while** statement in it.

```
<?xml version="1.0"?>
<!DOCTYPE wml PUBLIC "-//WAPFORUM//DTD WML 1.1//EN"
"http://www.wapforum.org/DTD/wml_1.1.xml">

<wml>
<card id="card1" title="While" newcontext="true">
<do type="accept" label="Get Result">
<go href="wmlex516.wmls#whilestmt()"/>
</do>
<p>
<u> $(result) </u>
</p>
</card>

</wml>

*** WMLScript File ***

extern function whilestmt() {

var i = 10;
var j = 0;
var answer = "";

while (i > j) {
 answer += i + " is more than " + j + ". \n";
 j++;
}
WMLBrowser.setVar("result",answer);
WMLBrowser.refresh();
}
```

Example 5.16

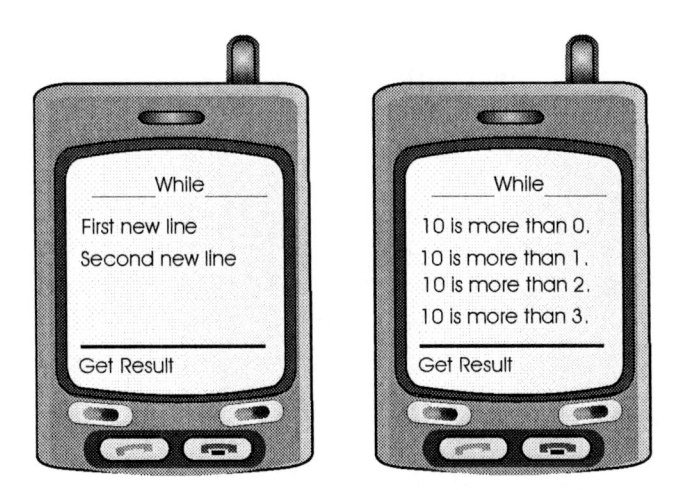

Figure 5.4 While Example

A for loop works as it does in C, Perl, or Java as a stepwise loop, where the initial value, condition, and increment (in that order) are specified within brackets and separated by semi-colons.

The initial value is normally used when initializing a counter type variable for iteration loops.

The condition in the **for** loop is an expression which returns a true or false. As long as the condition returns true, the loop's program block is executed. When the condition returns a false value, the program execution skips to the code following the loop program block.

---

for loop syntax

for (initial value; condition; increment)
statements

Example 5.17

Wireless Markup Language (WML)

The increment is used normally to change the initial value, by that increment, until a false condition occurs in the condition. Example 5.18 shows a **for** statement in a WMLScript program, called from a WML program. Note that in this example the variable i is declared within the **for** statement.

```
<?xml version="1.0"?>
<!DOCTYPE wml PUBLIC "-//WAPFORUM//DTD WML 1.1//EN"
"http://www.wapforum.org/DTD/wml_1.1.xml">

<wml>
<card id="card1" title="For">
<do type="accept" label="Get Result">
<go href="wmlex518.wmls#forstmt()"/>
</do>
<p>
<u> $(result) </u>
</p>
</card>

</wml>


*** WMLScript File ***

extern function forstmt() {

var answer = "";

for (var i=1; i <= 5; i++){

 answer += "Round " + i + " .\n";

}
WMLBrowser.setVar("result",answer);
WMLBrowser.refresh();
}
```

Example 5.18

Chapter 5

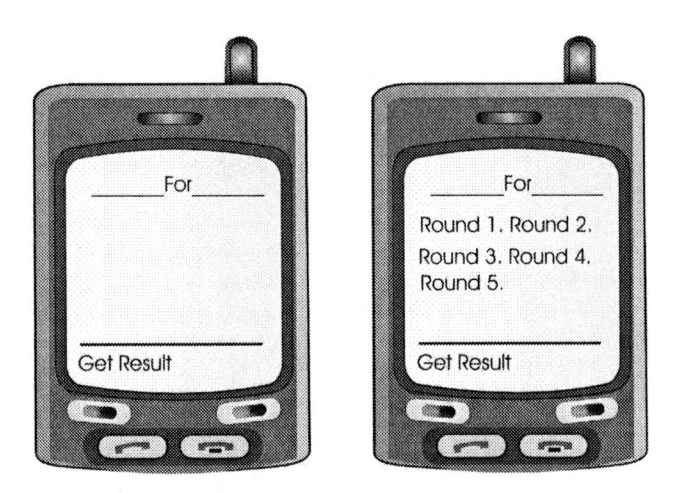

Figure 5.5 For Example

WMLScript also defines the statements **break** and **continue**. There is no **goto** statement in WMLScript. The **break** statement an only be used within a **while** or a **for** loop. When the program execution comes to a **break** statement in the block of code, it will immediately begin with the statement that follows the loop block., In other words, the **break** statement "breaks" the program execution out of the **while** or **for** loop, to the code following the loop. Example 5.19 shows the use of the **break** statement.

Wireless Markup Language (WML)

```
<?xml version="1.0"?>
<!DOCTYPE wml PUBLIC "-//WAPFORUM//DTD WML 1.1//EN"
"http://www.wapforum.org/DTD/wml_1.1.xml">

<wml>
<card id="card1" title="Break">
<do type="accept" label="Get Result">
<go href="wmlex519.wmls#dobreak($(number))"/>
</do>
<p>

Number (-5...5):
<br/>
<input type="text" name="number" title="Number"/>
<br/>

<u> $(result) </u>
</p>
</card>

</wml>


extern function dobreak(number) {


var answer = "";

for (var i=-5; i <= 5; i++){

 if (i == number) break;

}
 answer = i * number;
WMLBrowser.setVar("result",answer);
WMLBrowser.refresh();
```

Example 5.19

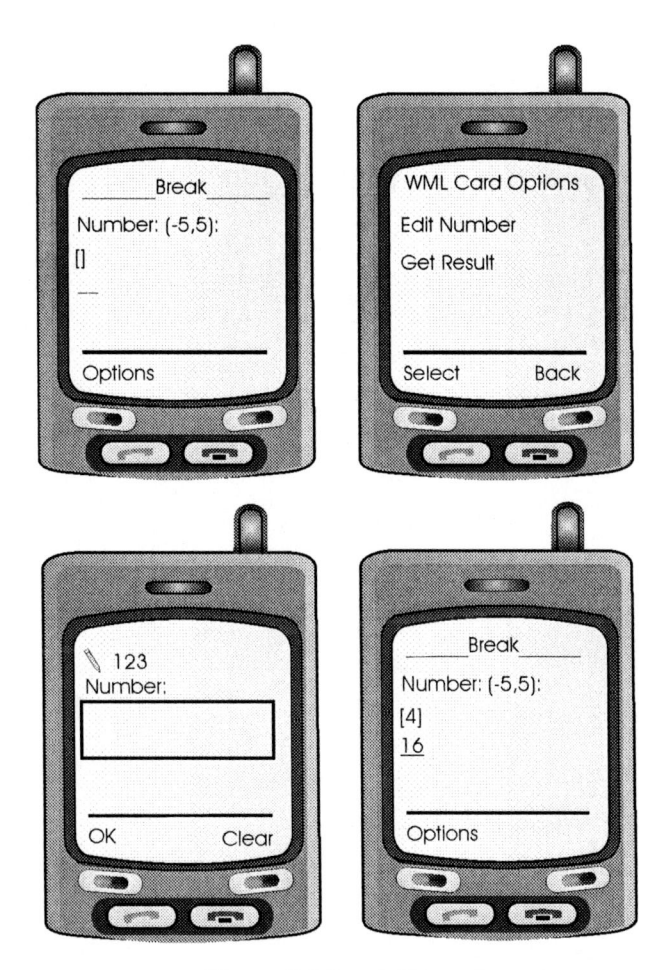

Figure 5.6 Break Example

The **continue** statement is used in a loop to move to the next iteration of that loop. By using the **continue** statement, you can interrupt a pass and start the next one in a **while** or **for** statement. The **continue** can only be used within loops. Example 5.20 shows the use of a **continue** statement within a **while** loop.

Wireless Markup Language (WML)

```
<?xml version="1.0"?>
<!DOCTYPE wml PUBLIC "-//WAPFORUM//DTD WML 1.1//EN"
"http://www.wapforum.org/DTD/wml_1.1.xml">

<wml>
<card id="card1" title="Continue">
<do type="accept" label="Get Result">
<go href="wmlex520.wmls#contn()"/>
</do>
<p>

<u> $(result) </u>
</p>
</card>

</wml>



*** WMLScript File ***

extern function contn() {


var answer = "";
 var i = 0;

while (i < 10){
  ++i;
 if (i%2 ==0) continue;
 answer += " * " + i;
}

WMLBrowser.setVar("result",answer);
WMLBrowser.refresh();
```

Example 5.20

136

Figure 5.7 Continue Statement

Calling WMLScript Functions

WMLScript code is written in normal text files with the extension *.wmls(i.e. ex51.wmls). These text files are placed on the same server as normal WML documents; however, it is also possible to call WMLScript functions on other servers from within WML documents or from WMLScript code. A text file is called a compilation unit, and it may contain one or several functions. A compilation unit can be placed in the same directory as WML documents, or in a separate subdirectory. Functions that have similar properties or belong to the same document should be placed in the same compilation unit.

When a WML document contains a reference to a WMLScript function, the call will be routed from the browser through the gateway to the server. The server will then send the necessary WMLScript compilation unit, which is converted into binary format in the gateway. This is done because the binary file is smaller and therefore easier to transmit over a wireless network. The binary file is sent from the gateway to the WAP browser. The WAP browser has a interpreter that is able to execute WMLScript programs in their binary format.

Wireless Markup Language (WML)

The same reference system is used for calling WMLScript functions as in HTML and WWW, which means that resources are referred to with their relative or absolute URL address. One URL address is always linked to one WMLScript compilation unit. The call should always include the name of the target

WMLScript function and its list of parameters. At the calling end of the function (in the WML document) the function call must have the same list of parameters that are in the WMLScript function. The call therefore needs to include the same number of parameters and in the same order as in the WMLScript function.

The WMLScript function may call functions within the same compilation unit, or functions in other compilation units or libraries. If the WMLScript function is defined with the word **extern**, it can be called from any other compilation unit, unless it is protected by access control programs.

A WMLScript compilation unit can protect its content by using a special access control pragma, which is examined before any externally-defined functions are executed. A compilation unit may use only one access control pragma.

An access control pragma is specified by the unique word **access**. The level of protection is always checked before the function is executed. If the compilation unit contains the defined reserved words **use access** in front of the domain name (with the reserved word **domain**) and/or a path (with the reserved word **path**), the authority of the compilation unit that calls the pragma-protected function to execute the function will be checked before execution.

In example 5.21, access is allowed to all locations in the *wap* directory (and its subdirectories) in the *wap.acta.fi* domain.

Use of the **access** pragma

use access domain "wap.acta.fi" path "/wap";

Example 5.21

Chapter 5

Summary

- WMLScript is a object-oriented, procedural scripting language whose purpose is to enhance the abilities WML.

- In WMLScript variables are not defined during declaration, but during execution.

- WMLScript supports local standard libraries – Lang, Float, String, URL, WMLBrowser, and Dialogs.

- When a WML document calls a WMLScript function, the parameter list of the calling document must match the parameter list in the WMLScript function, exactly.

- Variable names are strings of unlimited length that start with a letter or an underscore, and continue with letters, digits, or underscores.

- The command **var** declares a variable in WMLScript code.

- In WMLScript, the assignment operator is the equals sign (=).

- WMLScript allows the use of either short assignment operations, or full-length assignment operations.

- WMLScript allows bit shift operations, such as shift to left, shift to right, and shift to right(sign to positive value).

- Value addition operators, such as ++, —, which increase or decrease the value of the operand by 1, are allowed in WMLScript.

- Logical operators – AND, OR, and NOT are allowed in WMLScript.

139

Wireless Markup Language (WML)

- Comparison operators exist in WMLScript such as ==, !=, >, <, >=, <=.

- WMLScript also includes the operator **isvalid.**

- WMLScript does not include arrays.

- WMLScript has an **if-else** structure.

- A case statement is not included in WMLScript.

- WMLScript also has a **while** and a **for** statement.

- WMLScript has a **break** and a **continue** statement that must be used within a **while** or a **for** block of code.

- WMLScript code is written in normal text files with the extension *.wmls, and is stored on a server.

- WMLScript function may call functions within the same compilation unit, or functions in other compilation units or libraries.

Chapter 5

Questions

1.) What is the main difference between Javascript and WMLScript?

2.) Name the six standard libraries that WMLScript supports.

3.) What are the four types of values that can be declared for WMLScript variables?

4.) Which of the following variables are invalid names: var3, _num3, 4digit, VAR7?

5.) How can variables be passed to a WMLScript function, if it isn't listed in the parameter list on the calling statement or the function call?

6.) How would you write the WMLScript statement for dividing a variable (z) by 3?

7.) List the logical operators that WMLScript supports.

8.) Can **if-else** statements be nested in WMLScript?

9.) What is the comparison operator for greater than or equal to?

10.) Explain how the operator **isvalid** works.

11.) How does the **while** and the **for** statements differ?

12.) What does the **continue** statement do?

13.) Where physically, are the WMLScript programs stored?

14.) What does **use access** do?

Wireless Markup Language (WML)

Problems

1.) Write a WML program and WMLScript function that gets the user to enter a number, and then passes this number to a WMLScript function that multiplies that number by 5, and divides it by 2, and then returns the result.

2.) Write a WML program and WMLScript function that has the user enter a number, and then the WMLScript function checks to see if the number is between 5 and 8. If it is, the function should return a string "the number is between 5 and 8" or if not, return a string " the number is not between 5 and 8".

3.) Write a WML program and WMLScript function that returns a string of all the even numbers between 0 and a number entered by the user.

4.) Write a WML program and WMLScript function that has the user enter a social security number. The function should check to see if it is a match with a pre-stored SSN, and if it is, return to a card in the deck with the title and word "successful Match" on it. If it doesn't match, return to a card in the deck with the title and words " Not a Match."

5.) Write a WML program and WMLScript function that has a user define a password on one card, and then go to another card to enter the password, and if they match – return a "Match", if they don't match, return a "Non-match."

Chapter 6

Advanced WML Script

The WMLScript language supports automatic type conversion of variables, and it also supports a number of standard libraries with standard functions in them.

Type Conversions

In some cases, the operands of the WMLScript language require certain types of variables as their operands. In these cases WMLScript supports automatic type conversion, where the "wrong" types of variables are converted to match the data type required.

WMLScript is a weakly typed language, and the variable declarations do not specify a type. Internally the language handles the following data types: Boolean, integer, floating point, string and invalid. The **typeof** operator determines the internal data type of a variable.

Each WMLScript operator accepts a predefined set of operand types. If the provided operands are not of the right data type, an automatic type conversion must take place.

Integer and floating-point values can be converted to strings, which correspond to their original formats. The integer 5 can be represented as the string "5", and the floating-point number 0.5 as either ".5", "0.5", or ".5e0". Also, a Boolean value will also retain its original format, so the value **true** becomes the string "true" and **false** becomes the string "false".

Wireless Markup Language (WML)

A string can be converted to an integer value only if it contains a decimal representation of an integer number. "6" will convert to the integer 6, but "Bill6" will not convert to an integer. In some situations a conversion from string to integer may cause an error, in which case the conversion becomes **invalid**. Floating-point values cannot be converted to integers. But, the truth value **true** converts to the integer value 1, and **false** to 0.

A string can be converted to a floating-point value if it contains a floating-point representation. ".7" and "0.7" convert to 0.7, but "B0.7" does not convert to a floating-point value. If the conversion from string to floating-point causes an error, the conversion becomes invalid.

Integers convert to their corresponding floating-point values, so 4 becomes 4.0. And **true** converts to 1.0 and **false** to 0.0.

An empty string ("") converts to **false**, and all other strings are converted to **true**. This can be used when testing in a condition whether a string is empty or not. An integer value of 0 converts to **false**, and all other integers convert to **true**. A floating-point value 0.0 converts to **false**, but all other floating-point values convert to true.

WMLScript operators use the appropriate conversion rules when they are required. The conversion rules of many operands are specified in steps, whereby the operand conversion takes place before the operation, since the operation determines the expected operand type. If the operand types are compatible in the same operation, no conversion takes place.

WMLScript has a few "strong" operations, which only return certain value types and only accept certain operand types. These operators are: !, &&, || - the logical operators, which only accept truth values and only returns a truth value. Operators which return an integer value and only accept integer value operators are : -, <<, >>, >>>, &, ^, |, %, and **div**.

Other WMLScript operators can accept different types of variables as operands for an operation. The operators **typeof** and **isvalid** accept operands of any type. **Typeof** returns an integer type value, and **isvalid** returns a truth value.

Chapter 6

Example 6.1 shows a WML program calling a WMLScript which has several type conversions in it. Figure 6-1 shows the screens for example 6.1

```
<?xml version="1.0"?>
<!DOCTYPE wml PUBLIC "-//WAPFORUM//DTD WML 1.1//EN"
"http://www.wapforum.org/DTD/wml_1.1.xml">

<wml>
<card id="card1" title="Conversions">

<do type="accept" label="Get Result">

<go href="wmlex61.wmls#conversion()"/>
</do>
<p>
<u>$(result)</u>
</p>
</card>

</wml>
```

Example 6.1

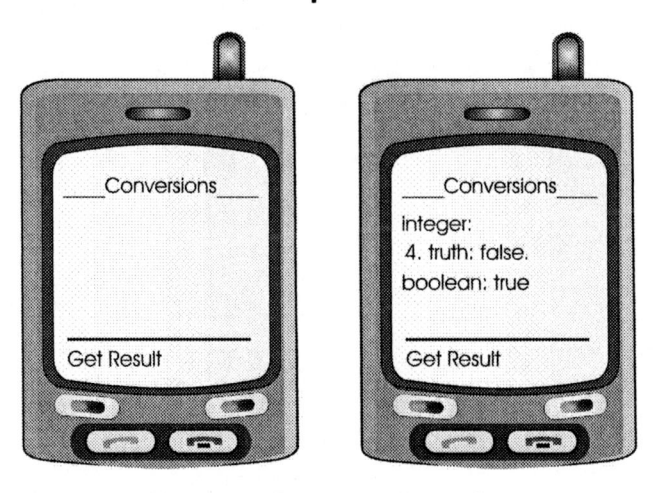

Figure 6.1 Conversion Example

Wireless Markup Language (WML)

Standard Libraries

WMLScript contains several standard libraries with functions that can be called by other functions in their own compilation units. Libraries are collections of functions that are logically grouped together based on what the library functions do. The functions are called by using a dot separator, as has been shown in examples in the last two chapters. This is similar to other object-oriented languages with the library name to the left of the dot, and the called function and its parameters to the right.

These libraries are stored in the WAP browser, which is resident in the WAP-compliant device. There are currently eight libraries available to use in WMLScript: **Lang, Float, String, URL, WMLBrowser, Dialogs, Debug,** and **Console**.

Example 6.2 shows a WMLScript function, which uses the **length** function from the **String** library, and the **setVar** and **refresh** functions from the WMLBrowser library. The **length** function returns the length of the string, and the **setVar** function sets the value of a variable, and **refresh** updates the WML document in the browser.

Figure 6.2 shows the associated screens.

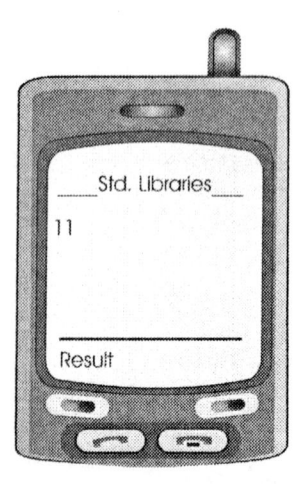

Figure 6.2 Library Example1

```
<?xml version="1.0"?>
<!DOCTYPE wml PUBLIC "-//WAPFORUM//DTD WML 1.1//EN"
"http://www.wapforum.org/DTD/wml_1.1.xml">

<wml>
<card id="card1" title="Std Libraries">

<do type="accept" label="Result">

<go href="wmlex62.wmls#chars()"/>
</do>
<p>
<u>$(result)</u>
</p>
</card>

</wml>
```

WMLScript File

```
extern function chars() {

var strngname="LibraryCard";

var answer = String.length(strngname);
// length function from String library returns
// a value of the length of the string

// functions from WMLBrowser library

WMLBrowser.setVar ("result", answer);
WMLBrowser.refresh();

}
```

Example 6.2

Wireless Markup Language (WML)

Appendix II lists all of the functions in each of the WMLScript libraries. The following sections will describe the purpose of each library, and give an example of one or more of the functions in that library. But, see Appendix II for all of the functions and what they do.

String

The String library contains functions for handling strings. In WMLScript a string is essentially treated as an array of characters, with the first element at offset of 0. The String library contains 16 functions, all which are listed in Appendix II.

The **charAt** function requires a string and an integer as parameters. It basically returns a string of the character from the input string, at the location specified by the integer. Example 6.3 shows the use of the **charAt** function. Figure 6.3 shows the associated screen.

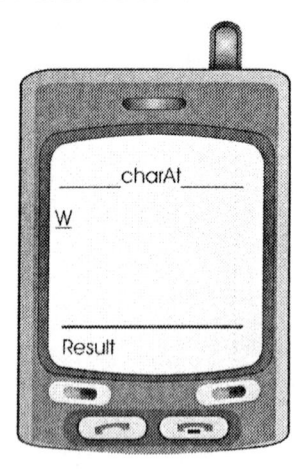

Figure 6.3 charAt Example

```
<?xml version="1.0"?>
<!DOCTYPE wml PUBLIC "-//WAPFORUM//DTD WML 1.1//EN"
"http://www.wapforum.org/DTD/wml_1.1.xml">

<wml>
<card id="card1" title="charAt">

<do type="accept" label="Result">

<go href="wmlex63.wmls#whichchar()"/>
</do>
<p>
<u>$(result)</u>
</p>
</card>

</wml>
```

*** WMLScript File ***

```
extern function whichchar() {

var str1 = "Tiger Woods is a world class champion";

var answer = String.charAt(str1,6);

WMLBrowser.setVar ("result", answer);
WMLBrowser.refresh();
}
```

Example 6.3

Wireless Markup Language (WML)

The **find** function from the String library is used to find from the first parameter string the part that is indicated by the second parameter string. If the part string is found in the target string the function returns an integer, which represents the index of the first character in the second parameter string. If the given partial string is not found, the function returns the value −1.

Figure 6.4 shows the screen associated of example 6.4.

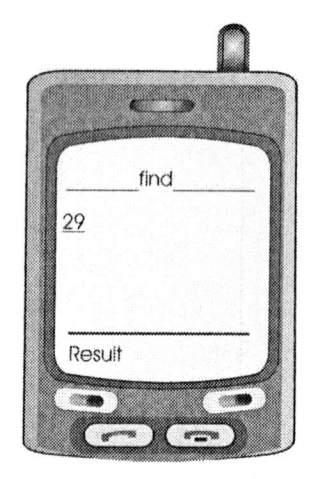

Figure 6.4 Find Example

Example 6.4 shows the use of the **find** function.

```
<?xml version="1.0"?>
<!DOCTYPE wml PUBLIC "-//WAPFORUM//DTD WML 1.1//EN"
"http://www.wapforum.org/DTD/wml_1.1.xml">

<wml>
<card id="card1" title="find">

<do type="accept" label="Result">

<go href="wmlex64.wmls#findit()"/>
</do>
<p>
<u>$(result)</u>
</p>
</card>

</wml>
 *** WMLScript File ***

 extern function findit() {

 var str1 = "Tiger Woods is a world class champion";

 var a = String.find(str1,"champion");

 WMLBrowser.setVar ("result", a);
 WMLBrowser.refresh();

 }
```

Example 6.4

Wireless Markup Language (WML)

The **isEmpty** function returns the value of true, if the length of the parameter string is 0, in other cases it returns the value false.

Example 6.5 shows this function, and figure 6.5 shows the associated screen.

The other functions in the String library are covered in Appendix II.

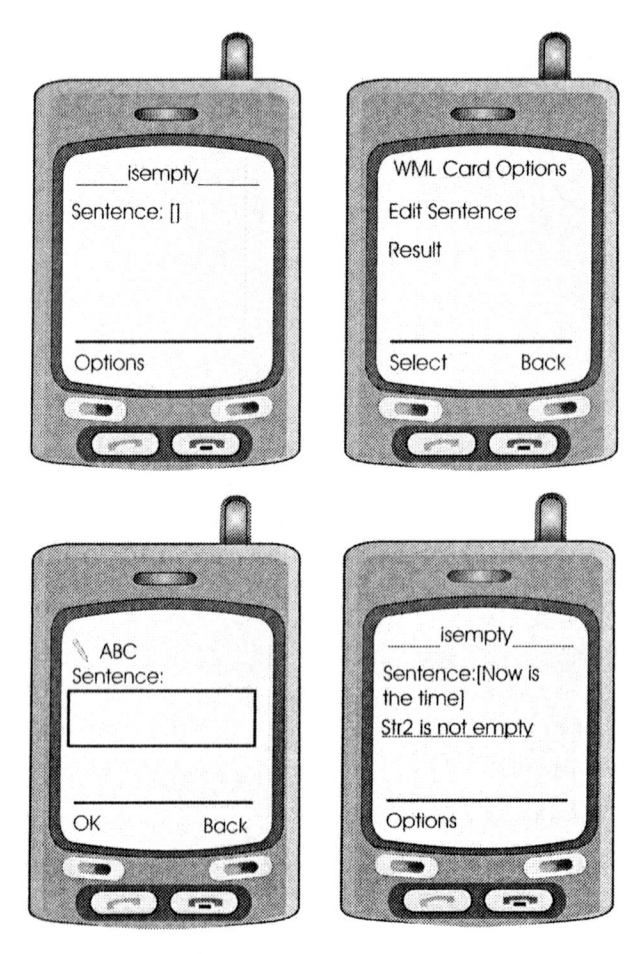

Figure 6.5 String Example2

```
<?xml version="1.0"?>
<!DOCTYPE wml PUBLIC "-//WAPFORUM//DTD WML 1.1//EN"
"http://www.wapforum.org/DTD/wml_1.1.xml">

<wml>
<card id="card1" title="isempty">

<do type="accept" label="Result">

<go href="wmlex65.wmls#empty('$(str2:unesc)')"/>
</do>
<p>
Sentence:
<input name="str2" title="Sentence" format="*m"/>
<br/>
<u>$(result)</u>
</p>
</card>

</wml>
  extern function empty(str2) {


  var answer;

  if (String.isEmpty(str2)){
   answer = "Str2 is empty";
  }
  else {
   answer = "Str2 is not empty";
  }

  WMLBrowser.setVar("result", answer);
  WMLBrowser.refresh();

  }
```

Example 6.5

Wireless Markup Language (WML)

Lang

The Lang library has functions that have a significant link to the WMLScript language. Library functions can be used to check variable types, imply type conversions, and influence the execution of the script.

There are 15 functions in the Lang library. They are all listed in Appendix II.

The **abs** function of the Lang library returns the absolute value of a parameter value. The parameter value can be either an integer or a floating-point value, the function returns the same type of value. Example 6.6 shows the use of the **abs** function. Figure 6.6 shows the associated screen.

```
<?xml version="1.0"?>
<!DOCTYPE wml PUBLIC "-//WAPFORUM//DTD WML 1.1//EN"
"http://www.wapforum.org/DTD/wml_1.1.xml">

<wml>
<card id="card1" title="Absolute">

<do type="accept" label="Result">

<go href="wmlex66.wmls#abs()"/>
</do>
<p>

<u>$(result)</u>
</p>
</card>

</wml>
```

Example 6.6

Chapter 6

```
extern function abs() {


var a = -10;
var b = 20;
var c = -0.5;
var d = 2.5;
var answer="";

var e = Lang.abs(a);

answer += e + ",";
 e = Lang.abs(b);
answer += e + ",";
 e = Lang.abs(c);
answer += e + ",";
 e = Lang.abs(d);
answer += e + ",";

WMLBrowser.setVar("result", answer);
WMLBrowser.refresh();

}
```

Example 6.6 (cont.)

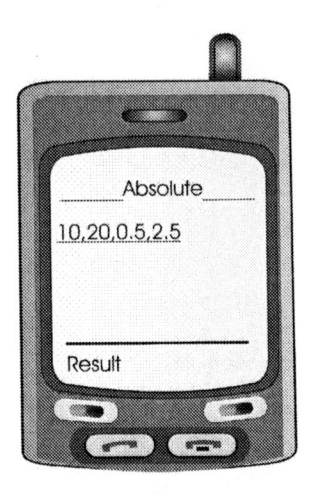

Figure 6.6 abs Example

Wireless Markup Language (WML)

The **isFloat** function returns the value true if it is able to convert a parameter to floating-point by using the **parseFloat** function. Strings can be converted to floating-point values if they have a floating-point presentation format.

Integers are converted directly to their corresponding floating-point values. (e.g. the integer 5 becomes the floating-point value 5.0.). If a parameter cannot be converted to a floating-point value, the function returns the value of false.

Example 6.7 shows the use of the **isFloat** function in the Lang library. The Figure 6.7 shows the associated screen.

All of the functions in the Lang library are listed in Appendix II.

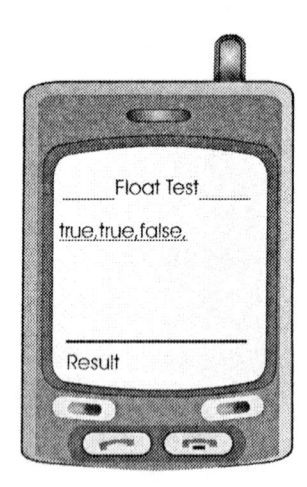

Figure 6.7 IsFloat Example

```
<?xml version="1.0" encoding="utf-8"?>
<!DOCTYPE wml PUBLIC "-//WAPFORUM//DTD WML 1.3//EN"
 "http://www.wapforum.org/DTD/wml13.dtd">
<wml>

 <card id="card1" title="Float Test">
<do type="accept" label="result">
<go href="wmlex67.wmls#isitfloat()"/>
</do>
<p>
<u> $(result)</u>
</p>
</card>
</wml>
*** WMLScript File ***

var a=Lang.isFloat("2000");

answer += a + ", ";

 a = Lang.isFloat("09.10");

answer += a + ", ";

a = Lang.isFloat("string");

answer += a + ", ";

WMLBrowser.setVar ("result",answer);
WMLBrowser.refresh();
}
```

Example 6.7

Float

The Float library routines deal with the processing of floating-point values. The library contains the most typical functions that are needed to execute floating-point calculations, but also calculations involving integer values.

All of the functions in the Float library are listed in Appendix II, but for an example of one, consider the **pow** function. This **pow** function from the Float library returns the floating-point value which is the value of the first parameter (base) to the power of the second parameter (exponent). Example 6.8 shows this function. Figure 6.8 shows the associated screen.

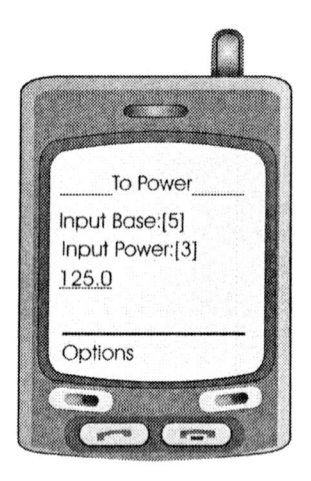

Figure 6.8 POW Example

```
<?xml version="1.0" encoding="utf-8"?>
<!DOCTYPE wml PUBLIC "-//WAPFORUM//DTD WML 1.3//EN"
 "http://www.wapforum.org/DTD/wml13.dtd">
<wml>

 <card id="card1" title="To Power">
<do type="accept" label="result">
<go href="wmlex68.wmls#tothepower($(num1),$(num2))"/>
</do>
<p>
Input base:
<input type="text" name="num1" title="Base"/>
<br/>
Input power:
<input type="text" name ="num2" title="Power"/>
<u> $(result)</u>
</p>
</card>
</wml>


*** WMLScript File***

extern function tothepower(num1,num2) {

var answer;

answer = Float.pow(num1,num2);

WMLBrowser.setVar ("result",answer);
WMLBrowser.refresh();

}
```

Example 6.8

Wireless Markup Language (WML)

The **round** function is another Float library function. It returns an integer that is the closest approximation of the entered integer or floating-point parameter value. If the parameter is an integer, it will be returned as such. If the parameter is a floating-point value, the round function will use common mathematical approximation rules when converting it to the closet possible integer. Example 6.9 shows the use of the **round** function.

```
<?xml version="1.0" encoding="utf-8"?>
<!DOCTYPE wml PUBLIC "-//WAPFORUM//DTD WML 1.3//EN"
 "http://www.wapforum.org/DTD/wml13.dtd">
<wml>

 <card id="card1" title="Rounding">
<do type="accept" label="result">
<go href="wmlex69.wmls#rounding($(num))"/>
</do>
<p>
Input No. to be Rounded:
<input type="text" name="num" title="Number"/>
<br/>

<u> $(result)</u>
</p>
</card>
</wml>

*** WMLScript File ***

extern function rounding(num) {

var answer;

answer = Float.round(num);


WMLBrowser.setVar ("result",answer);
WMLBrowser.refresh();
}
```

Example 6.9

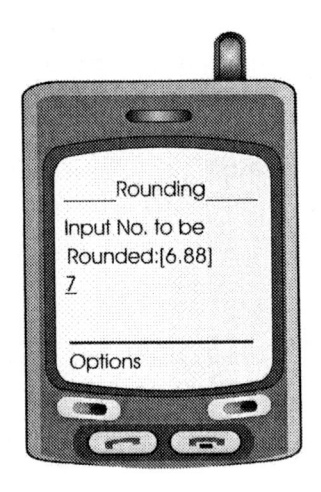

Figure 6.9 Round Example

URL

The URL library contains functions that can be used to examine validation, content and parameters of absolute and relative URL addresses. The library also includes functions for searching the contents of new URL addresses.

The **escapestring** function from the URL library returns from the entered parameter a new string that is URL encoded. This means that all special characters have been changed to hexadecimal format. This method is used when sending form data to CGI programs and Java servlets using the POST or GET methods. Example 6.10 uses the **escapestring** function. The figure shows the associated screen

Wireless Markup Language (WML)

```xml
<?xml version="1.0" encoding="utf-8"?>
<!DOCTYPE wml PUBLIC "-//WAPFORUM//DTD WML 1.3//EN"
 "http://www.wapforum.org/DTD/wml13.dtd">
<wml>

 <card id="card1" title="Escape String">
<do type="accept" label="result">
<go href="wmlex610.wmls#escape('$(str)')"/>
</do>
<p>
Input a Test String:
<input type="text" name="str" title="String"/>
<br/>

<u> $(result)</u>
</p>
</card>
</wml>
```

*** WMLScript File***

```
extern function escape(str) {

var answer;

answer = URL.escapeString(str);

WMLBrowser.setVar ("result",answer);
WMLBrowser.refresh();

}
```

Example 6.10

Chapter 6

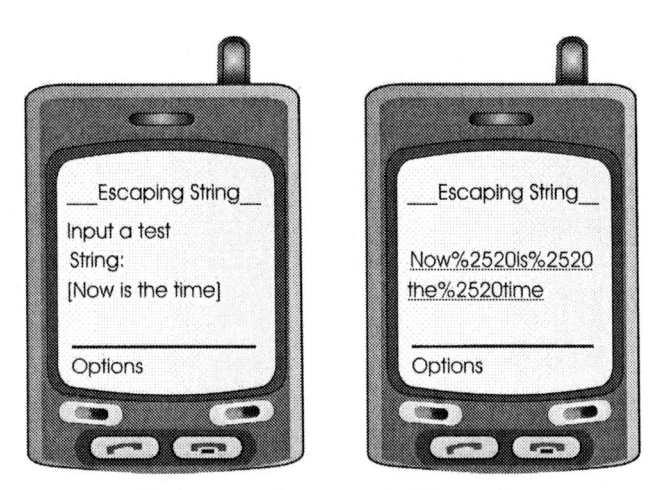

Figure 6.10 Escapestring Example

Just a note on inputting variables from a WML program to a WMLScript function: you need single quotes around the variable (e.g. '$(str)') if it is a string, but you don't need the single quotes around the variable if it is an integer or floating-point.

The **getbase** function from the URL library returns the absolute address of the WMLScript compilation unit. Example 6.11 shows the use of the **getbase** function. Figure 6.11 shows the associated screen

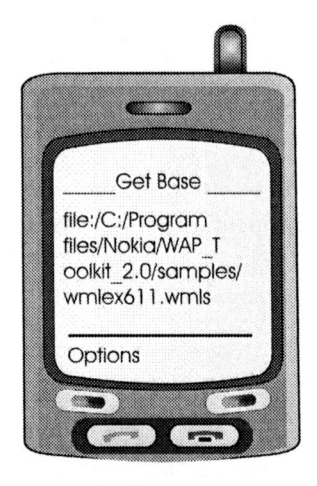

Figure 6.11 GetBase Example

163

Wireless Markup Language (WML)

```
<?xml version="1.0" encoding="utf-8"?>
<!DOCTYPE wml PUBLIC "-//WAPFORUM//DTD WML 1.3//EN"
    "http://www.wapforum.org/DTD/wml13.dtd">
<wml>

<card id="card1" title="Get Base">
<do type="accept" label="result">
<go href="wmlex611.wmls#thebase()"/>
</do>
 <p>

<u> $(result)</u>
</p>
</card>
</wml>

*** WMLScript File***

extern function thebase() {
 var a;
   a = URL.getBase();
WMLBrowser.setVar ("result",a);
WMLBrowser.refresh();
}
```

Example 6.11

WML Browser

The WMLBrowser library functions have been used in most of the examples in Chapters 3, 4, 5, and 6. The library contains functions which make it possible to examine WML document variables, to change their values and to manipulate the transfer between documents in a browser.

A variable that is defined in a WML document can be retrieved into a WMLScript function by using the **getVar** function from the WMLBrowser library. The variable name used in the WML document is the same variable name used in the WMLScript compilation unit. Basically this is another way to "pass" variables from WML documents and WMLScript functions, without listing them as parameters in the function call.

Example 6.12 shows the use of the **getVar** function. Figure 6.12 shows the associated screen.

```
<?xml version="1.0" encoding="utf-8"?>
<!DOCTYPE wml PUBLIC "-//WAPFORUM//DTD WML 1.3//EN"
 "http://www.wapforum.org/DTD/wml13.dtd">
<wml>

 <card id="card1" title="Get Var">
<do type="accept" label="result">
<go href="wmlex612.wmls#nopass()"/>
</do>
<p>
Enter Name:
<input type="text" name="name"/>

<u> $(result)</u>
</p>
</card>
</wml>

*** WMLScript File ***

extern function nopass() {
 var answer;
 answer = WMLBrowser.getVar("name");
WMLBrowser.setVar ("result",answer);
WMLBrowser.refresh();
}
```

Example 6.12

Wireless Markup Language (WML)

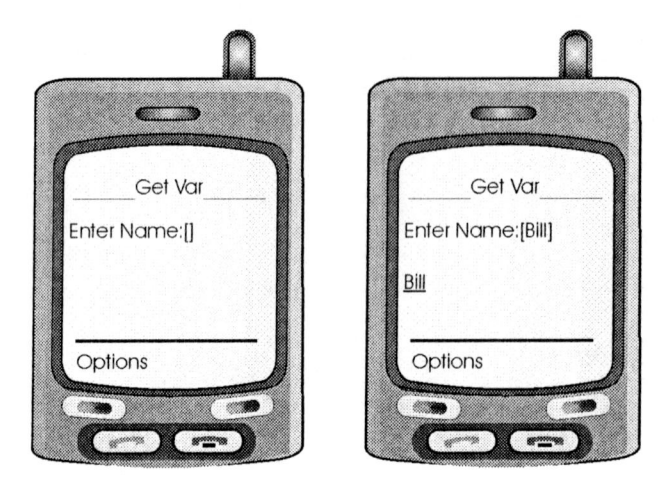

Figure 6.12 getVar Example

The **go** function in the WMLBrowser library defines the URL address where the browser moves when the control of the program execution transfers from the WMLScript interpreter back to the browser. Normally the go function is called as the last WMLscript command.

Example 6.13 shows the use of the **go** function as it appears in a WMLScript function.

```
extern function nopass() {
 var addr;
 addr = "wmlex613.wmls#funct1";

 WMLBrowser.go(addr);
}
```

Example 6.13

Chapter 6

The **setVar** and **refresh** functions from the WMLBrowser library are used in most of the examples in previous chapters. They set the value of variables (**setVar**) and refresh the contents of the browser(**refresh**).

The remaining functions of the WMLBrowser library are listed in Appendix II.

Dialogs

The Dialogs library in WMLScript has useful functions, which can be used to quickly create interactive dialog windows for the user. The library also contains functions that display user entries in the dialog window.

The **alert** function from the Dialogs library creates a message on the browser display, and expects the user to respond to it by pushing the ok button.

Example 6.14 shows the use of the alert function. Figure 6.14 shows the associated screen.

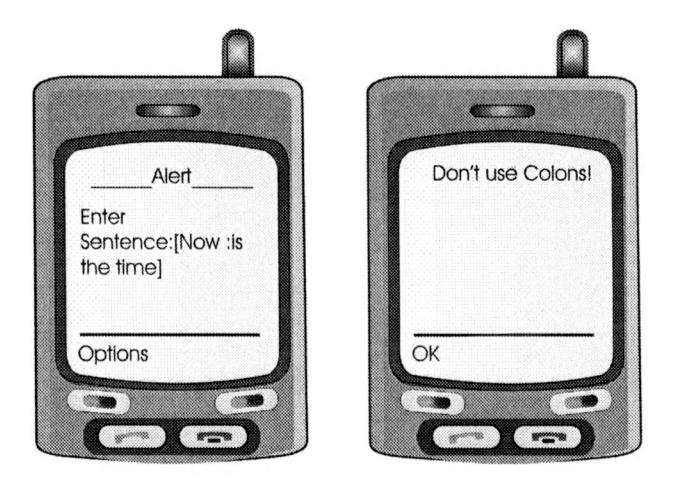

Figure 6.14 Alert Function

167

```
<?xml version="1.0" encoding="utf-8"?>
<!DOCTYPE wml PUBLIC "-//WAPFORUM//DTD WML 1.3//EN"
 "http://www.wapforum.org/DTD/wml13.dtd">
<wml>

 <card id="card1" title="Alert">
 <do type="accept" label="Dialog">
 <go href="wmlex614.wmls#dialog('$(sent:unesc)')"/>
 </do>
 <p>
Enter Sentence:
<input type="text" name="sent"/><br/>

</p>
</card>
</wml>

*** WMLScript File***

extern function dialog(sent) {

for (var i = 0; i < String.length(sent); i++){

 if (String.charAt(sent,i) == ":"){
 Dialogs.alert("Don't use Colons!");
 return;
 }
 }

}
```

Example 6.14

The **confirm** function from the Dialogs library creates a message on the browser screen and expects the user to respond before continuing the program execution. Example 6.15 shows the use of the confirm functions. Figure 6.15 shows the associated screen.

```
<?xml version="1.0" encoding="utf-8"?>
<!DOCTYPE wml PUBLIC "-//WAPFORUM//DTD WML 1.3//EN"
 "http://www.wapforum.org/DTD/wml13.dtd">
<wml>

 <card id="card1" title="Dialog">
<do type="accept" label="Dialog">
<go href="wmlex615.wmls#dialog()"/>
</do>
<p>
Opening Dialog window....$(test)

</p>
</card>
</wml>


*** WMLScript File***

extern function dialog() {

 var docontinue = Dialogs.confirm("Continue?","Yes","No");

 if (docontinue) {
 WMLBrowser.setVar("test","we continued");
 }
 else {
 WMLBrowser.setVar("test","we didn't continue");
 }
 WMLBrowser.refresh();

}
```

Example 6.15

The remaining functions in the Dialogs library are listed in Appendix II.

Wireless Markup Language (WML)

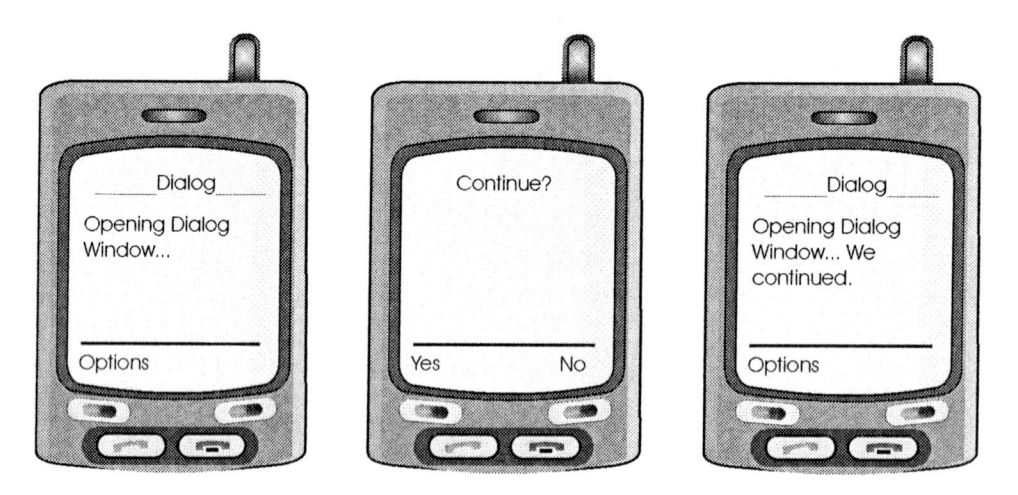

Figure 6.15 Continue Function

Debug

The Debug library is a Nokia-specific library which contains debugging functions. The functions in this library are for closing and opening files and printing – all to be used while debugging.

Console

The Console library is a Phone.com specific library containing debugging functions. Both functions do printing.

See Appendix II for details on each function in the Console Library.

Chapter 6

Summary

- WMLScript supports automatic type conversion of variables.

- WMLScript has a few "strong" operations which only return certain value types and only accept certain operand types. Other WMLScript operators can accept different types of variables as operands for an operation.

- WMLScript contains several standard libraries which are logical groups of functions in each library. The current libraries are: Lang, Float, String, URL, WMLBrowser, Dialogs, Debug, and Console.

- The String library contains functions for handling strings.

- The Lang library has functions that have a significant link to the WMLScript language. These functions can be used to check variable types, imply type conversions, and influence the execution of the script.

- The Float library functions deal with the processing of floating-point values. The library contains the most typical functions that are needed to execute floating-point calculations, but also calculations involving integer values.

- The URL library contains functions that can be used to examine validation, content and parameters of absolute and relative URL addresses. The library also includes functions for searching the contents of new URL addresses.

- The WMLBrowser library contains functions which make it possible to examine WML document variables, to change their values and to manipulate the transfer between documents in a browser.

Wireless Markup Language (WML)

- The Dialogs library in WMLScript has useful functions, which can be used to quickly create interactive dialog windows for the user. The library also contains functions that display user entries in the dialog window.

- The Debug library is a Nokia-specific library which contains debugging functions. The functions in this library are for closing and opening files and printing – all to be used while debugging.

- The Console library is a Phone.com specific library containing debugging functions. Both functions do printing.

Chapter 6

Questions

1.) What is meant by "automatic type conversion" of variables?

2.) What integer values will the strings : "8", "Fred5", and "eight" convert to?

3.) What Boolean value does an empty string convert to?

4.) List those WMLScript operators that only accept truth values and only return a truth value.

5.) Which operators accept operands of any type?

6.) List the eight standard libraries that WMLScript supports.

7.) Where do the standard libraries for WMLScript reside?

8.) What does the **charAt** function in the String library do?

9.) What do the Lang library functions basically cover?

10.) What does the following function do: Float.pow(3,2)?

11.) What library contains functions that are used to examine validation, content, and parameters of URL addresses?

12.) When is the escapestring function most often used?

13.) How can you get variables assigned in a WML document in a WMLScript function if they aren't passed as parameters in the function call?

14.) What does the **alert** function from the Dialogs library do?

15.) Are any of the libraries manufacturer-specific?

Wireless Markup Language (WML)

Problems

1.) Write a WML document that gets a sentence as input, and passes it to a WMLScript function that determines the length of the string and returns that length.

2.) Write a WML program which gets a string inputted, and then goes to a WMLScript program to see if the word "master" is in it. Return appropriate responses to indicate whether it does or doesn't have the word "master" in it.

3.) Write a WML program and a WMLScript program that has the user enter a number and then the WMLScript function returns both the absolute value as well as the rounded value of the number inputted.

4.) Write a WML program and a WMLScript program that gets an escaped string as input and converts back to an unescaped string and returns it.

5.) Write a WML program and a WMLScript program that has the user input two numbers – a base number and a power number – and then goes to a WMLScript function to calculate the base number to the power number – but pass no variables as parameters in the function call.

Chapter 7

Setting up a WAP Server

A WAP server is a software program that resides on a computer on the Internet, that acts as a WAP gateway and a WAP content server. This is the program that handles requests from wireless devices for WAP documents that reside on that server.

There are two ways to get a WAP server that will handle requests for WAP documents (written in WML or XHTML). One way is to use a server that is specifically designed for WAP, such as the Nokia WAP Server. Another way is to modify an HTTP server, such as the Apache HTTP server, to handle requests for WAP documents.

Figure 7.1 shows a WAP device requesting data through the Internet to a WAP Server. The WAP device requests a WAP content document through the Internet. The Internet sends the request to a WAP server for processing. The WAP server then sends back the requested data to the Internet, which in turn sends it to the actual WAP device that displays the data.

Wireless Markup Language (WML)

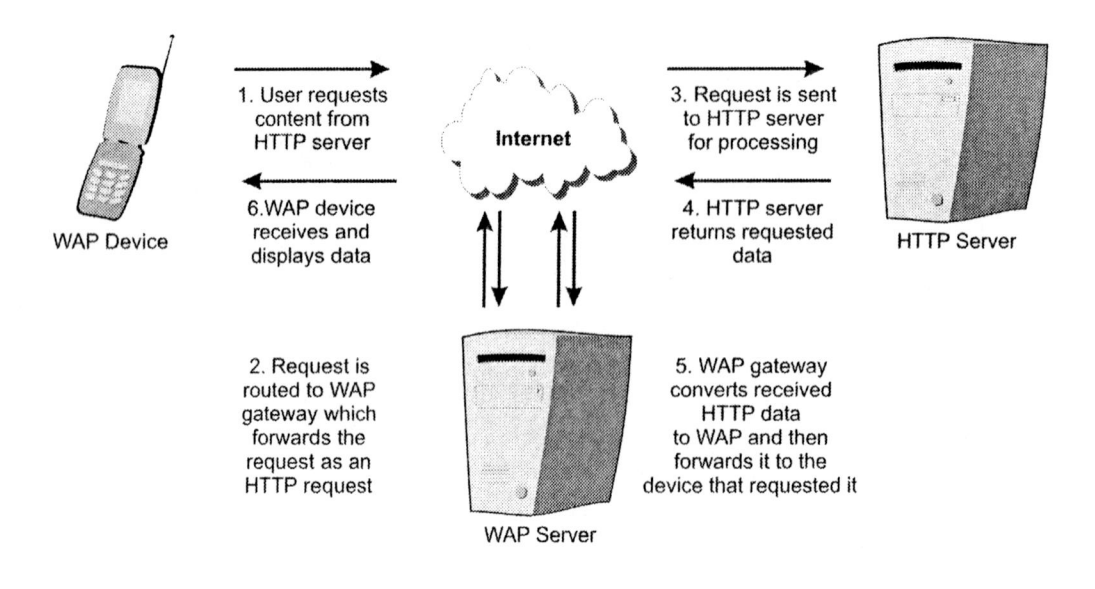

Figure 7.1 WAP Server

Nokia WAP Server

The Nokia WAP Server can be used as a WAP gateway server as well as a content server. The Nokia WAP Server performs protocol conversion between WAP and HTTP (or similar) protocols. It also provides support for user administration, authentication, administration of service access control for users and user groups, application development using Java Servlet API, etc. The server supports network interfaces for GSM data calls via the UDP/IP (User Datagram Protocol/Internet Protocol) bearer adaptor and GSM-SMS messaging (Global System for Mobile Communication, Short Message Service).

In addition to handling Java Servlets, the Nokia WAP Server can act as a content server for various kinds of applications and documents required by the WAP service. With the help of a graphic user interface, the administra-

Chapter 7

tor of the server is able to monitor the state of the server, statistics, log data and alarms, administer access levels, install, start and stop bearer adaptors and install new servlets and associate them with URL addresses.

Modifying an HTTP Server

Another way to get a WAP server is to modify an HTTP server so that it can handle WAP requests. It must be modified, since just putting your WAP applications up on a standard HTTP server will not allow users to view content on their phones. The server must tell the WML microbrowser that it is about to receive a WML page and not an HTML page or some other kind of content. This communication is done using MIME extensions that must be added to your server for WML microbrowsers to correctly read your content.

MIME is Multipurpose Internet Main Extension, which is a piece of header information that was originally used in email to allow for proper formatting of non-ASCII messages over the Internet. There are many predefined MIME types in common use, such as JPG graphics files and HTML files. In addition to email programs, Web browsers also support a variety of MIME types. This permits the browser to display or output files in formats other than HTML. MIME types common to Internet servers include:

- "text/html" for HTML files

- "image/jpg" for JPG files

- "image/gif" for GIF files.

Because these file types are so common, most Web servers are already configured to send the correct MIME types. WAP, however, requires its own MIME types to recognize various file content. By adding these MIME types to your server, different devices will be able to properly interpret and therefore display WAP content.

Wireless Markup Language (WML)

WAP currently requires five MIME types shown in Table 7.1 to serve WML, WMLScript, and Wireless Bitmap images.

File Extension	Content Type	MIME Type
wml	WML source code	text/vnd.wap.wml
wmls	WMLScript source code	text/vnd.wap.wmlscript
wbmp	Wireless Bitmaps	image/vnd.wap.wbmp
wmlc	Compiled WML	application/vnd.wap.wmlc
wmlsc	Compiled WMLScript	application/vnd.wap.wmlscriptc

Table 7.1 File Types and Corresponding MIME Types

Publishing to the WAP Server

WAP documents and applications should be thoroughly tested before they are published to the world. Testing is best done with an emulator, such as the Nokia WAP Toolkit. Testing should also be done, if possible, using several different WAP devices and browsers, to make sure the documents look acceptable, at least on the most commonly used browsers.

Once document functionality has been thoroughly tested on one's own computer, the documents can be transferred to an origin WAP server for the world to view. This transfer is done using either a character-based or a graphical FTP program. One of the most popular and easy to use graphical FTP programs is WS_FTP . This program allows you to send your files up to the WAP server, via the FTP protocol. You must send all code files – WML and WMLScript, as well as any image files that your documents reference. Currently all WAP browsers recognize WBMP format for images since this is part of the WAP standard. Some WAP browsers also recognize GIF and JPG formats, but usually mobile phones only accept WBMP format images.

Chapter 7

Summary

- A WAP server is a software program that resides on a computer on the internet, that acts as a WAP gateway and a WAP content server.

- There are two ways to get a WAP server: use a server that is specifically designed for WAP, or modify an HTTP server to handle WAP documents.

- The Nokia WAP Server performs protocol conversion between WAP and HTTP (or similar) protocols.

- The administrator of the Nokia WAP server is able to monitor the state of the server, statistics, log data and alarms, administer access levels, install, start and stop bearer adaptors and install new servlets and associate them with URL addresses.

- To modify an HTTP Server to handle WAP documents, you must add certain MIME extensions to your server.

- MIME types permit the browser to display or output files in formats other than HTML.

- By adding certain MIME types to your server, different devices will be able to properly interpret and therefore display WAP content.

- WAP documents and applications should be thoroughly tested before they are published to the world.

- Documents can be transferred to the WAP server via FTP, either character-based or graphical based.

- Currently all WAP browsers recognize WBMP format for images since this is part of the WAP standard.

Wireless Markup Language (WML)

Questions

1.) What is a WAP server?

2.) What is another way of "making" a WAP server?

3.) What functions can the administrator of the Nokia WAP server do?

4.) What is a MIME extension?

5.) List three common MIME extensions that exist in most Internet servers.

6.) What are the five MIME types that are necessary for WAP documents?

7.) What should be done before publishing your WAP document to a WAP Server?

8.) What protocol is commonly used to send your WAP files to a server?

Chapter 8

Creating Push and Pull Notifications

Notifications are events that are generated from the server, to contact the user. There are two kinds of notifications, push and pull.

What are Notifications

A notification is data delivered to a client, whose delivery is initiated by the server. There are many common terms used to describe notifications, such as asynchronous message delivery, content push, push messaging, or mobile and active channels.

First, consider the opposite model to notifications – a synchronous model that is most commonly used in Internet data delivery. On the Web, client requests information from a server, and the server responds with the appropriate file. In the asynchronous model, (the one used in notifications), the servers don't have to wait for a request to deliver data. Instead, the server uses a set of rules to determine when it should send the data to a particular client, and does so as soon as possible.

When would you use notifications in a WML application? For one, you might want to alert your users to severe weather or traffic conditions. Or, you might want to let a group of users know when tickets to their favorite music group go on sale, and then send them a link to purchase tickets. For these kinds of applications, it is necessary to have a central application server generate an event that affects the WAP device's display or operation.

Wireless Markup Language

Although the WAP Forum has created a specification for notifications, not all WAP gateways support notifications. Because they are generated on the server-side of an application, the WAP server running the application must support them.

In order to work with notifications, you will need to know the subscriber ID of the person you are sending the message to. A subscriber ID uniquely describes a WAP user to the WAP gateway. Each device has a subscriber ID, and no two are alike. That information can be pulled through WML to identify the target device uniquely. Secondly, you need access to a network that supports notifications.

There are essentially four distinct types of messages or notification that can be pushed to a WAP device: alerts, cache operations, content, and multi-part messages.

Alerts

Alerts contain a brief message and a URL. When an alert is triggered, the device may bring it to the user's attention in several different ways. The device may beep, buzz, ring, vibrate, or do something else.

For example, you could program an alert to notify the user if a certain stock fell below a certain price. If that happened, the server could "push" a message to the WAP device saying the stock fell and a URL to his WAP stock service. The user could take the link to his stock service or not.

Alerts are very useful for "for your information" notifications. Alerts may take up to a half an hour or longer to be delivered, and therefore they should never be used for critical data or data that requires an immediate response. The short message and URL format is optimal for keeping the message size down; if the user wants additional information, he can use the URL contained in the alert.

Chapter 8

Cache Operations

The second types of notifications you can create are cache operations. Unlike alerts, users do not see any evidence that a cache operation has occurred on their device. Cache operations are an "invisible" kind of notification and are best used to deal with application variance.

An example might be when a food menu has been created on a WAP site, and certain customers receive a discount on some menu items. In order to minimize download times, you might have their WAP devices regularly cache the menu and price list. When a price changes, the easiest way to make that change is to tell their WAP device to eliminate the cache for the menu deck. In this case, a command from the server will invalidate a deck or set of decks for a single device's cache. This is very useful when we work with data that changes.

When a cache control notification occurs, cached data is just invalidated; no actual data will be transmitted at that time.

Content Messages

The third type of notification delivers standard WAP content. This type of message can deliver any sort of WAP-based content such as a WML deck, a WBMP image, or a WMLScript function. There are some interesting applications for content-push messages, yet most WAP devices do not currently support them. In many cases, a standard alert notification will serve as a workable substitute for sending an entire file or image to a WAP device.

Multi-part Messages

The last kind of notification to be considered is a multi-part message. Multi-part messages contain multiple types of notification contents and are sometimes referred to as a digest. A multi-part message may contain an alert, a cache operation, and a content message, or it may contain any combination of two types of notifications. For example, if we updated our stock prices once a day, it would be useful to clear the user's cache, but also alert them that new data has arrived. Not all browsers support multi-part messaging at this time, but Phone.com's UP browser does.

Push versus Pull Notifications

There are two actual methods for pushing messages down to a WAP device, push and pull. Before this chapter we have always thought of a model in which our WAP device requests a file from a server, and that server then delivers the data. This is called the pull method. The push method, on the other hand, uses the server to actually send data to the WAP device without any kind of request.

A pull notification relies on the WAP device to poll the server for new events and information. Therefore, based upon a device's availability the message might wait at the WAP gateway until the message can be processed.

Figure 8.1 shows how push notification works with a WAP server. This example shows a WAP push proxy gateway that receives a message from a stock news information service that sends a message via the Internet when to a WAP client when a particular stock changes beyond a present limit. When the WAP push proxy gateway receives the message, it immediately alerts the WAP client of an incoming message. Because the WAP client has been setup to automatically receive these pushed messages, the WAP client acknowledges the request from the push sever and the message is delivered (pulled) from the push proxy gateway to the WAP client. This example shows that the information can be automatically displayed on the WAP client without the user having to do anything.

Chapter 8

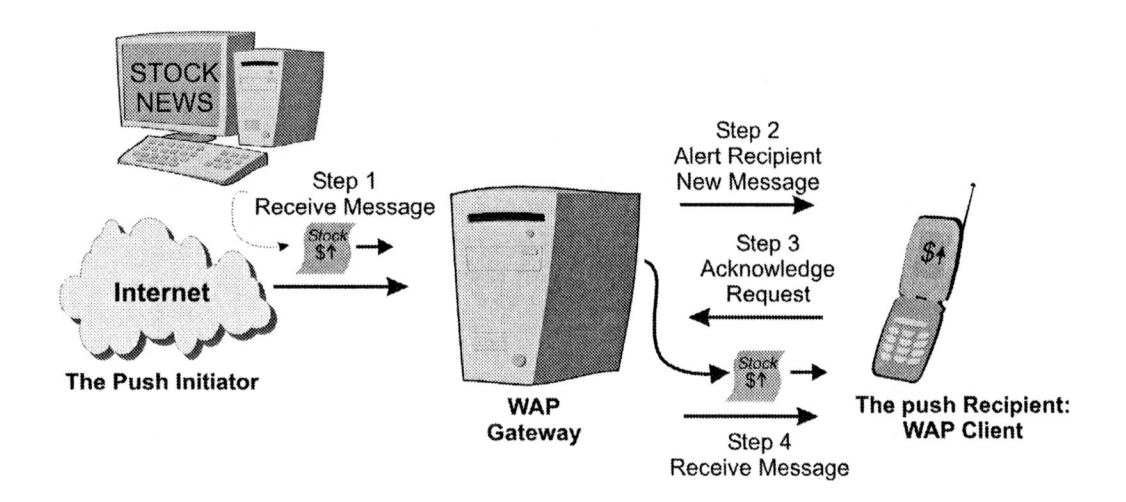

Figure 8.1 Model for Push Notification

The following steps for notification processing are shown in figure 8.1.

A notification is sent to the gateway for delivery to the user.

If the user's handset is turned on, the notification is delivered. If the notification is an alert, the user is notified by a sound and/or a message. If the notification is a cache operation, it is performed without any visual or audible cues. If the user's handset is turned off, the notification is queued at the server until it is delivered, or until the TTL (Time-To Live) has expired. After the user receives the notification, he can choose to view the URL associated with it or not.

Wireless Markup Language

If the user chooses to fetch the URL, the UP.link server requests the content at the URL.

The content is then sent back to the WAP gateway and delivered by the WAP gateway to the device.

A pull notification is sent in a similar fashion.

Figure 8.2 shows how pull notification works with a WAP server. This example shows a WAP push proxy gateway that receives email messages that are addressed to the WAP client. The push proxy gateway stores these messages until it receives a request from the WAP client for the delivery of messages. The WAP client will then download (pull) the messages from the push proxy gateway so the messages can be displayed on the users phone.

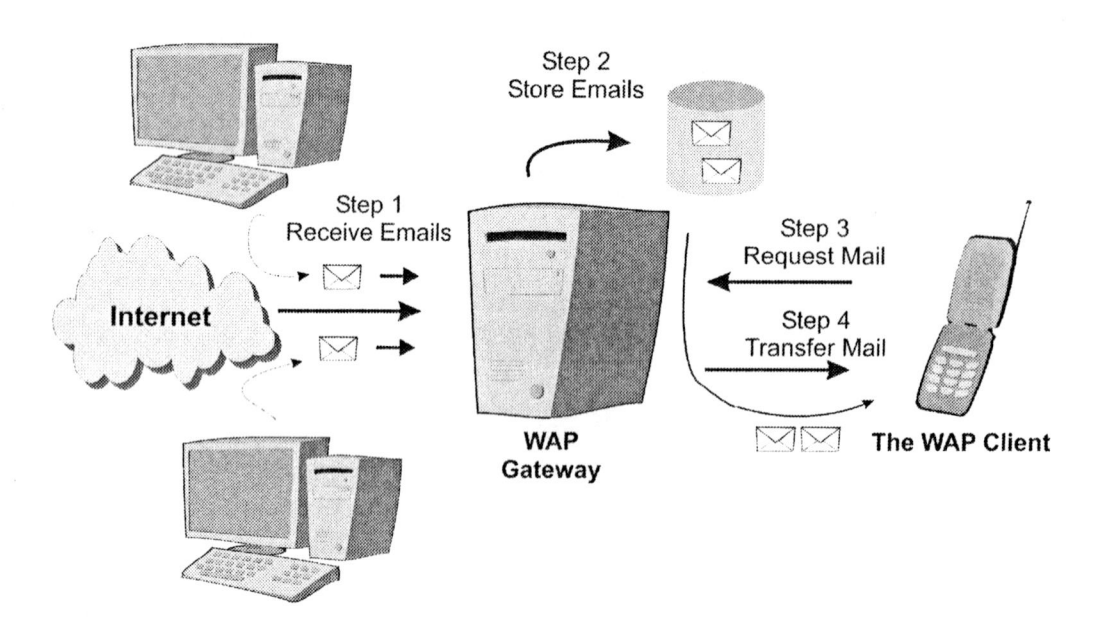

Figure 8.2 Model of a Pull Notification

Chapter 8

The following steps for notification processing are illustrated in figure 8.2.

A notification is sent to the WAP gateway for delivery to the user.

The WAP gateway holds the notification in the queue until the client requests it. (This assumes the client is on a circuit-switched network. For packet-switched networks, the notification will be delivered immediately.)

The next time the user's device is turned on and a connection is made to the network, and pending notifications is requested from the gateway.

The WAP gateway delivers the notification in the form of a digest to the user's device. You can include an alert entry in the notification digest, which will visually and/or audibly signal the arrival of a new notification to the user.

Pull notifications are constantly executed until one of the two conditions occurs:

The message is successfully delivered.

The Time-To-Live (TTL) on the message expires, and the message sent fails to be delivered.

Typically, you will most often use push notifications.

Summary

- Notifications are events that are generated from the server, to contact the user.

- A notification is data delivered to a client, whose delivery is initiated by the server.

Wireless Markup Language

- Although the WAP Forum has created a specification for notifications, not all WAP gateways support notifications. Because they are generated on the server-side of an application, the WAP server running the application must support them.

- In order to work with notifications, you will need to know the subscriber ID of the person you are sending the message to.

- There are four distinct types of messages or notification that can be pushed to a WAP device: alerts, cache operations, content, and multipart messages.

- Alerts contain a brief message and a URL.

- Cache operations are an "invisible" kind of notification and are best used to deal with application variance.

- A content message can deliver any sort of WAP-based content such as a WML deck, a WBMP image, or a WMLScript function.

- Multi-part messages contain multiple types of notification contents and are sometimes referred to as a digest.

- There are two actual methods for pushing messages down to a WAP device, push and pull.

- A pull notification relies on the WAP device to poll the server for new events and information.

- The push method uses the server to actually send data to the WAP device without any kind of request.

Chapter 8

Questions

1.) Describe the asynchronous model of getting information to the user.

2.) List five possible applications that could use notifications.

3.) What is the subscriber ID?

4.) Name the four types of messages or notifications that can be pushed to a WAP device.

5.) What does an alert notification contain?

6.) What type of notification is essentially invisible to the user?

7.) What is the current problem with content message notifications?

8.) What are multi-part message notifications?

9.) What are the two methods for pushing messages to a WAP device?

10.) What are the two conditions that will end a pull notification?

Wireless Markup Language

Chapter 9

Adding Security to Applications

Creating a WML application usually means that you will be sending and receiving data across the Internet as the user takes his browser with him from place to place. There are inherent insecurities in this environment and WAP has provided some facilities to address these insecurities. You want to make sure that when you create WML applications they can be trusted to handle sensitive information and even cash.

Security Basics

Your WAP application is basically an Internet application. In order for a user to reach your application, he must connect through his local Internet service provider (ISP) to a WAP gateway before requesting data from your Web server. This request from the WML browser will travel between many different systems before it reaches your application code.

Any application on the Internet is subject to some basic security problems.

Figure 9.1 shows some of the methods that may be used to compromise the security of messages that are sent between a sender and a receiver through the Internet. This diagram shows that someone could be eavesdropping (copying and redirecting) the message. Someone could be intercepting (capture) and retransmitting the message. An unauthorized user could impersonate the receiver information by using the receiver's own identification and authentication information to control a communication session.

Wireless Markup Language (WML)

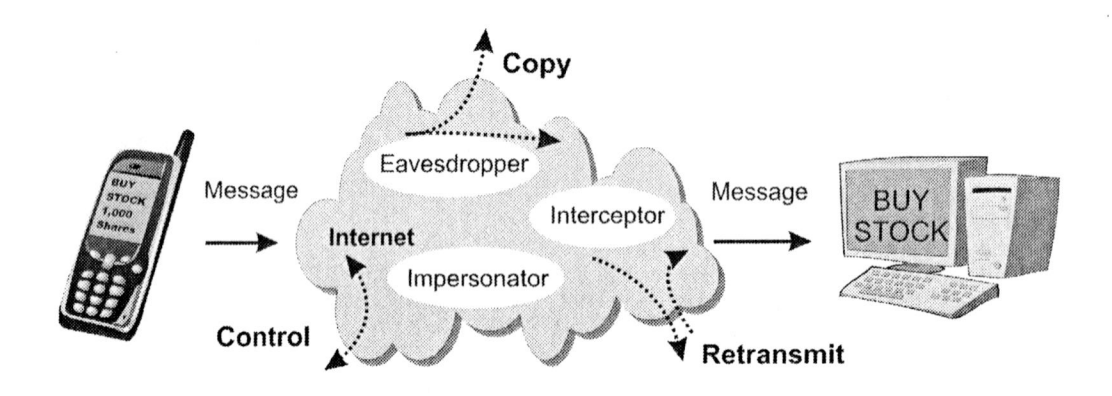

Figure 9.1 Internet Message Security Issues

In figure 9.1, the sender might be a user, and the receiver would be your application. Internet security systems often provide confidentiality, authentication, integrity, and non-repudiation for data exchanged between the user and the application.

Confidentiality. It should be difficult for an eavesdropper to understand the contents of a message between the sender and receiver. Systems often achieve this goal by encrypting the message so that it would take many resources (time, computing power, and so on) to retrieve the original.

Authentication. Both the sender and receiver should be able to verify the identity of the other and reject impersonators. Systems often implement this using a trusted third party that vouches for the identity.

Integrity. The receiver of a message requires assurance that the contents received are identical to the contents sent. A security system typically accomplishes this by providing a signature for each message, based on its contents, that is very hard to alter.

Non-repudiation. Finally, either the sender or the receiver of the message must be permitted in the future to prove that the message was sent and received. This is often done by requiring each party to sign its message using a signature registered with a trusted third party.

Not every application security system addresses each of these issues; each will only provide the facilities that are practical and required for the environment.

Today, most widely deployed Web applications use HTML browsers that employ a technology called the *Secure Sockets Layer* (SSL) that ensures confidentiality. This communication layer sits above the HTTP and TCP/IP layers and encrypts each request and response to and from the server. To accomplish this encryption, SSL uses *public key cryptography*, in which both parties map plain text into cipher text. The information sender encrypts using a private value, or key, and gives the receiver a key to reverse the operation.

The SSL also provides integrity by adding a mathematical hash value to each message. It calculates this value by examining each piece of the message; the calculation algorithm guarantees that it is very difficult to replace any or the entire message and still produce the same hash.

In addition to SSL, most secured Web sites use a digital certificate purchased from a third party that vouches for the identity of the site's owner. This allows the browsing user to verify that the site to which he is sending data is actually the site he intends. This provides authentication of the receiver. Most sites use name and password to authenticate the sender because, although personal identification digital certificates are available, they are still not widely used.

Wireless Markup Language (WML)

In similar fashion to the authentication typically employed, most Web applications implement weak non-repudiation. In fact, most use only the authentication information they gather from the user to prove that he transmitted a message.

Threat Models

Now that you have an idea of what facilities a security system should provide your application, let's consider why it should provide them. Each piece of the security system exists to thwart a particular attack, or threat model. These models attempt to clarify how a foe might cause you and your users harm.

The first and probably most often cited threat is the *eavesdropper* attack. This threat is a passive attack in which the foe does not interfere with the exchange of information; he merely records the data for other, probably unauthorized, uses. The attacker simply watches the data travel from the sender to receiver and records a copy of it.

To prevent the eavesdropping attack, most security systems provide confidentiality using an encryption algorithm. Of course, no encryption is or ever will be perfectly secure; the eavesdropper, given enough time and energy, will be able to break through it. To be effective, a security system simply recognizes that and chooses a scheme that makes the price much higher than the value of the information exchanged.

Another threat model is the *impersonation* attack, sometimes known as the *Trojan horse* attack. This is a more active attack in which the attacker creates a harmful Web application that appears to the user to be legitimate. This rogue application can then ask the user for his authentication information or even sensitive data such as a credit card or social security number.

Effective Trojan horse applications will gather the desired data from the sender and pass it along to the legitimate application. To prevent this attack, the legitimate site contracts a trusted third party to vouch for its identity and provide a mechanism for clients to verify that identity when

Chapter 9

making requests. This, of course, requires the client to check these credentials before sending sensitive information.

The man-in-the-Middle attack is an active attack similar to the Trojan horse; the attacker places himself between the sender and receiver as shown in figure 9.2. This allows him to intercept all communication between the two parties and replace it with his own.

Figure 9.2 shows how an Internet security attacker can place him between the user and the application. As the man-in-the-middle, the attacker can receive the users request, and then alter it before sending it to the application. Then, the attacker can get the applications response, alter it, and send it back to the browser.

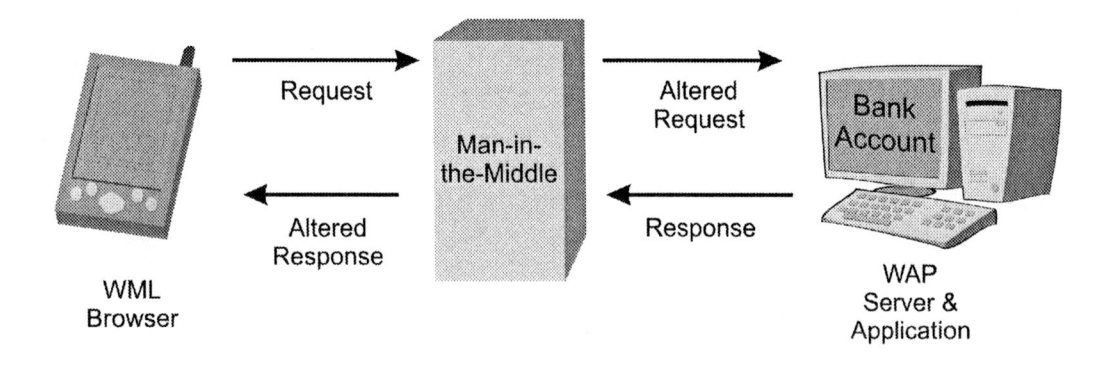

Figure 9.2 Man-in-the-Middle WML Security Attacker

Wireless Markup Language (WML)

Since the attacker has interposed him and both the sender and receiver think that he is the other party, he has both the opportunity to monitor all information exchanged and to alter that information. This is a potent threat that a security system can again counter with proper use of authentication.

A more direct threat to the security of your application lies in the *Dictionary* attack. In the direct variant of this attack, the foe takes a list of commonly used passwords and attempts to log into your application as a known user. There are several ways to reduce the threat against this happening to your application: you can require passwords that are hard to guess (for example, [word][number][word] format) or you can prevent the attacker from trying multiple successive attempts at authentication.

One final threat model with which you may find yourself faced while creating Web applications is the *Replay* attack. In this scenario, the attacker eavesdrops on a session between a user and your application. Once the user has stopped using the application, the attacker simulates another identical set of operations.

This threat is especially potent when, for example, the user is transferring funds. In order to prevent this, the security system will need to ensure that the authentication of the sender is time-sensitive and a later replay with the same data fails.

WAP Security Architecture

The WAP platform has leveraged the lessons from the development of other platforms, such as the HTML browser, and it has built-in facilities to address security.

Request Path

WAP requests travel from the wireless device through the wireless network to a WAP gateway. This gateway pushes the request on through the Internet to their ultimate destination – your Web Server.

Chapter 9

Figure 9.3 shows a simplified communication path between a WAP communication device (a mobile phone) and a WAP application server. This diagram shows that the WAP gateway assists in the setup and management of connections between the wireless network and WAP server by acting as a proxy (redirection) server. The WAP communication session request goes through the wireless network to the WAP Gateway The WAP gateway passes the communication session request to the application program residing on a WAP Server. Responses from the application program passes through the Internet to the WAP gateway, and finally back to the wireless device.

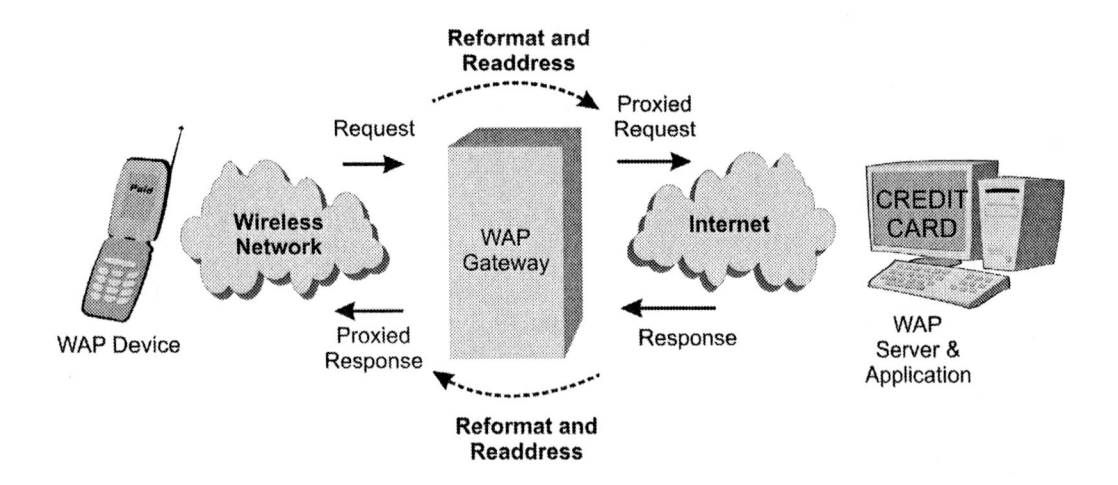

Figure 9.3 WAP Communication Session Establishment

Wireless Markup Language (WML)

On the left-hand side of the diagram in figure 9.3, you can see the wireless WAP device exchanging requests and responses with the WAP gateway. This communication takes place over the wireless network and its security is the exclusive responsibility of the WAP protocol stack.

At the WAP gateway, requests from WML browsers are translated into HTTP requests for data across the Internet. The WAP gateway then executes the request on behalf of the browser, constructing and passing along HTTP variables such as HTTP_REFERER.

Once the Web server processes the request, hands it to your application, and has a response to send, it replies to the WAP gateway. This communication path is no different than any other request/response cycle to a Web application; it's simply done on behalf of the browser rather than by the browser — it's a proxy request.

The WAP gateway then takes the WML in the HTTP response from the Web application and compiles it into the WAP Binary XML Content Format (WBXML). The gateway then completes the request by sending this WBXML version of the response to the wireless device.

This request/response path to retrieve a WML deck presents a few opportunities for attack. Here are some example attacks:

- Eavesdropping between the wireless device and the WAP gateway,

- Man-in-the-Middle at the WAP gateway,

- Eavesdropping between the WAP gateway and the Web server,

- Replay of the request to the WAP gateway.

The WAP specification attempts to ward off these threats by placing a transport layer security into its protocol. This transport layer security addresses the confidentiality and integrity requirements of the system.

Chapter 9

Like SSL, however, WAP transport layer security does not address authentication or non-repudiation. To address these concerns, you have to build your own mechanisms, just as you do in HTML applications today.

WTLS and SSL

The WAP platform provides you with facilities at the transport level to ensure confidentiality and message integrity. The Wireless Transport Layer Security (WTLS) provides security for the data exchanged directly between the wireless device and the WAP gateway. It ensures confidentiality and integrity in a very similar fashion to SSL, using public-key cryptography. In fact, WTLS is strongly based on the Transport Layer Security (TLS) 1.0 specification, which, in turn, derives from the SSL 3.0 specification.

The WTLS protocol, like its siblings, employs strong cryptographic algorithms such as RSA and RC5 (Rivest Cipher Five). The RSA algorithm is general purpose, but computationally intensive, whereas RC5 is lightweight enough while remaining secure to be employed by wireless devices. WTLS will typically use the former algorithm to exchange temporary keys used in the latter algorithm. This layer repels the eavesdropping and replay attacks on the wireless connection described earlier in this section.

Communication between the WAP gateway and your Web application must be secured using SSL in much the same fashion as you would secure it for an HTML application. In fact, since the WAP gateway is simply a proxy for the WML browser, the security of that part of the request/response chain has all of the features and problems of a secured HTML site.

Since you will need to use a secured Web server for your WAP application, all of your URLs will likely begin with https rather than http. This will repel any eavesdropping or replay attacks from the WAP gateway to your Web server.

There remains an attack on the security of your WAP application that has no WAP-specified protocol to defeat: the compromise of the WAP gateway. The gateway translates WTLS messages into SSL messages and vice versa.

Wireless Markup Language (WML)

Since this machine is literally the man-in-the-middle, were it to be hijacked by a foe, all of the conversations between wireless devices and your or other Web applications would be insecure.

While this is truly a hole in the WAP security model, there is only limited opportunity to compromise these gateways. Gateway providers must practice standard security procedures such as:

- Employing a firewall,

- Limiting administrative access to the machine to critical parties only,

- Limiting physical access to the machine,

- Only using gateway software that avoids persistent storage of plain text messages; in fact, using software that destroys the plain text as soon as possible,

- Automatically monitoring the machine for new process creation and other indications of compromise with professional management software.

With the WAP gateway secured, the platform provides excellent confidentiality and data integrity.

Security Certificates

One of the future improvements for the WAP security platform is the plan to have wide employment of client security certificates for authentication.

Key pairs for public-key cryptography are often very computationally intensive to create, far beyond the capabilities of typical WAP devices. In order to create the public/private key pair once for a client and present it to WAP gateways for securing the WTLS layer, WAP utilizes a security certificate.

Chapter 9

The WAP client security certificate is based on the ISO specification named X.509. At its most basic, the X.509 certificate incorporates a unique name for the client, a period of validity, the public key for the client, and the signature of a trusted third party vouching that the key is owned by the named client and valid.

A full-blown X.509 certificate is rather large and the WTLS provides a mini-certificate designed for the use on limited capability devices. This wireless device offers up this certificate to the WAP gateway during establishment of the WTLS layer. The gateway can then validate that the device is what it claims to be by checking that the signature from the trusted third party is intact.

The creation and deployment of these certificates is often referred to as the public-key infrastructure (PKI). The enhancement of PKI to support client authentication above the WTLS layer will complete the WAP security system since you will no longer need to insert your own authentication and non-repudiation into your applications.

Session Management

In order to tailor your WAP application to a user, you will need to create a software session in which you can store context information. This session may contain the name of the user and any data that he generates during his visit, such as items in a shopping cart.

WAP has no support for application sessions built in to its definition. You will have to manually create them by authenticating the user and tracking his progress. The next few sections will address how to do this, but it should be remembered that the WML applications should be as simple as possible to use – you should minimize the information burden you place on the user.

Wireless Markup Language (WML)

Client Authentication

There are two main ways to authenticate a user for your application: manually in a deck or with built-in Web server basic authentication. The former requires that you create a user database with login names and passwords, while the latter requires that you use a user database that your Web server can access.

To create a manual authentication deck, you simply need to ask the user for his name and password and then pass him on to the deck he wanted originally. This is a very simple process in most cases.

An alternative approach to creating your own authentication mechanism is the built-in facility in the Web server. This built-in facility is *Basic Authentication* – HTTP supplies it in IETF RPC 2617.

The Web browser and server implement the Basic Authentication scheme below the application layer. To understand how this works, let's look at the communication pattern between the browser and sever. The Web server generates a response code for each HTTP request that it sees. Often this response code is "OK", however, when the request is for a page that is protected, the response code tells the browser to challenge the user for his identity.

This identity challenge tells the browser, HTML or WML, to prompt the user for his username and password. The browser then sends this information back to the Web server immediately and the server attempts to verify the information. When the user supplies authentic information to the challenge, the Web server responds with the "OK" and the originally requested page. When the user fails to supply a valid combination of username and password, the Web server responds with the response code "Forbidden" and does not send the page.

In order to implement the Basic Authentication facility, the Web server requires a user directory. This user directory might be unique to your Web application or a shared corporate directory. Directory services vary widely

Chapter 9

between Web servers; Microsoft IIS integrates natively with the Microsoft Active Directory, while Netscape Enterprise Server uses Netscape Directory Server.

Once authentication is successful, each subsequent request from the browser to the Web server includes the username and password of the user. The Web server then creates an application variable for the Web application called REMOTE_USER containing the name of the user.

In addition to using a manual authentication structure or Basic Authentication, you may be able to take advantage of platform-specific identification information. For example, the WAP gateway that Phone.com delivers sends an HTTP variable that is unique to the wireless device making the request. This variable is named HTTP_X_UP_SUBNO and is tied directly to the device; the gateway constructs it from the wireless connection. You may be able to take advantage of this server variable in your application.

WML for Secure Applications

When creating WML decks, you have a few techniques available to ensure that sensitive information does not leave the secure communication channel between the user and your application. You can protect your decks from attacking applications using the **access** element and the **sendreferer** attribute of the **go** element.

WML allows you to tell the browser to keep other applications from linking into decks within your application. That is, WML provides you the **access** element in which you specify where the linking deck must live in order to reference your deck successfully.

You can add the **access** element to the <head> tag of any deck. The absence of the <access> tag allows any deck to reference that deck. When configuring the element, you should specify the **path** and **domain** attributes. The **domain** attribute controls the set of Web servers that can link into your application.

The **path** attribute restricts access within a domain in order to thwart cross-application linking. It permits any decks in a more specific path access.

The **sendreferer** attribute of the **go** element is quite simple to use – it can only be set to true or false. This attribute instructs the browser to send along the HTTP variable describing the deck that makes a request, when it is set to true. You can then decide, in you application logic, whether to permit the linking deck access by checking this variable. Unfortunately, the sendreferer attribute defaults to the false value, suppressing this valuable information. You should get into the habit of specifying this attribute as true each time you create a go element, even if you don't need access control immediately.

Cleaning Up

There is one last thing that you must take care of when you create an application that handles sensitive information: you must clean out your variables. This is critically important when you create decks that retrieve passwords, and social security numbers, for example. You cannot leave the data that the user enters into the browser in memory since the next deck may be able to access it and deliver it to another application against the user's wishes.

In order to clear out the variables before using them, you should use the **onenterforward** event, **refresh**, and **setvar** elements. You should make sure that each time you retrieve data from the user and leave the deck; you also clear out the in-memory data. You can simply do this with the **setvar** element at the exit point of the program setting all variables back to null. This is a good habit to get into, not only for sensitive data, for all data your deck uses.

Chapter 9

Summary

- Any application on the Internet is subject to some basic security problems.

- Internet security systems often provide confidentiality, authentication, integrity, and non-repudiation for data exchanged between the user and the application.

- *Confidentiality*. It should be difficult for an eavesdropper to understand the contents of a message between the sender and receiver.

- *Authentication*. Both the sender and receiver should be able to verify the identity of the other and reject impersonators.

- *Integrity*. The receiver of a message requires assurance that the contents received are identical to the contents sent.

- *Non-repudiation*. Finally, either the sender or the receiver of the message must be permitted in the future to prove that the message was sent and received.

- Most widely deployed Web applications use HTML browsers that employ a technology called the *Secure Sockets Layer* (SSL) that ensures confidentiality.

- SSL uses *public key cryptography*, in which both parties map plain text into cipher text.

- The SSL also provides integrity by adding a mathematical hash value to each message.

Wireless Markup Language (WML)

- The first and probably most often cited threat is the *eavesdropper* attack. This threat is a passive attack in which the foe does not interfere with the exchange of information; he merely records the data for other, probably unauthorized, uses.

- To prevent the eavesdropping attack, most security systems provide confidentiality using an encryption algorithm.

- Another threat model is the *impersonation* attack, sometimes known as the *Trojan horse* attack. This is a more active attack in which the attacker creates a harmful Web application that appears to the user to be legitimate.

- The man-in-the-Middle attack is an active attack similar to the Trojan horse; the attacker places himself between the sender and receiver, and intercepts all communications.

- A more direct threat to the security of your application lies in the *Dictionary* attack. The foe takes a list of commonly used passwords and attempts to log into your application as a known user.

- In the *Replay* attack, the attacker eavesdrops on a session between a user and your application. Once the user has stopped using the application, the attacker simulates another identical set of operations.

- The WAP specification attempts to ward off security threats by placing a transport layer security into its protocol. This transport layer security addresses the confidentiality and integrity requirements of the system.

- The Wireless Transport Layer Security (WTLS) provides security for the data exchanged directly between the wireless device and the WAP gateway. It ensures confidentiality and integrity in a very similar fashion to SSL, using public-key cryptography.

- Communication between the WAP gateway and your Web application must be secured using SSL in much the same fashion as you would secure it for an HTML application.

- One of the future improvements for the WAP security platform is the plan to have wide employment of client security certificates for authentication.

- There are two main ways to authenticate a user for your application: manually in a deck or with built-in Web server basic authentication.

- You can protect your decks from attacking applications using the **access** element and the **sendreferer** attribute of the **go** element.

- The **sendreferer** attribute instructs the browser to send along the HTTP variable describing the deck that makes a request, when it is set to true.

- There is one last thing that you must take care of when you create an application that handles sensitive information: you must clean out your variables.

Wireless Markup Language (WML)

Questions

1.) List four features that Internet security systems should provide.

2.) What is SSL?

3.) List five main security threats.

4.) What is the most common way of preventing an eavesdropping attack?

5.) How is the Trojan horse attack often prevented?

6.) Explain how the Man-in-the-Middle threat works.

7.) In a typical wireless Internet access, where are the most likely places for attacks?

8.) What is WTLS?

9.) What standard security procedures should gateway providers practice?

10.) What is a security certificate?

11.) What are the two main ways to authenticate a user for a WML application?

12.) Explain Basic Authentication.

13.) How do you clear your variables in a WML program?

Chapter 10

Other Script Languages

As discussed in Chapter 1, there currently are three main languages being used to program wireless devices. These are WML and WMLScript, XHTML Basic, and cHTML. WML and WMLScript have been covered in chapters 3-6, and XHTML Basic and cHTML will be covered in this chapter.

As of now, WML and WMLScript and XHTML Basic are the most widely used languages, with cHTML being used only in the I-mode service provider area. Currently WML and WMLScript and XHTML Basic are being consolidated by OMA in their WAP 2.0 Specifications. The cHTML has not been merged or consolidated yet into the WAP Specifications, but due to its similarity to XHTML Basic, this will probably happen in the future. In the mean time, both XHTML Basic and cHTML are covered in this chapter.

Both XHTML Basic and cHTML rely on the use of cgi-bin programs written in perl, java, or C++ to accomplish what WMLScript does for WML. Therefore, XHTML Basic and cHTML are languages to only develop the documents for the wireless devices browsers.

XHTML

XHTML Basic is the World Wide Web's Consortium's (W3C) initiative to provide a common markup language for wireless devices and other small devices with limited memory. The W3C released its recommendation for

Wireless Markup Language (WML)

XHTML Basic in December 2000. Also, WAP 2.0 Specification includes XHTML Basic with WML and WMLScript. WAP 2.0 basically allows WML and WMLScript to be embedded within XHTML Basic.

XHTML Basic is similar to WML in that it was derived from XML. XHTML Basic is a smaller version of XHTML, and therefore won't jeopardize the limited memory of wireless devices. XHTML Basic excludes features in XHTML that are not well-suited for wireless clients, such as frames, nested tables and nested tables.

XHTML Basic documents are created with a text editor such as Wordpad or Notepad, but are also included in the newer SDKs such as the Nokia Toolkit 4.0. Nokia Toolkit 4.0 can develop WML, WMLScript , and XHTML Basic documents. They are given the file extension xhtml, html, or htm. Also, XHTML Basic documents can be created in a text editor, and then displayed on a Pixo Internet Microbrowser.

XHTML Basic documents use start and end tags, like HTML and WML. The **html** element encloses both the head section (represented by using the **head** element) and the body section (represented by using the **body** element). The head section contains information about the XHTML Basic document, such as the document's title. The body section contains the content that a browser displays when users visit the Web page.

Example 10.1 shows a simple XHTML Basic document. The first three lines are required in XHTML Basic documents, and will be the same in all XHTML Basic documents. The fourth line has the <**html**> tag, which has a required attribute of **xmlns** and a value of http://www.w3.org/1999/xhtml.

Comments in XHTML Basic are delineated with <!— comment — >. This is the same as WML. Next, the <head> tag and the </head> tag surround the head section of the document. The <title> tag is located in the head section, and in this example simply has a title of "Welcome." This title generally shows up in the title bar of the browser. As in most markup languages, the text included within the start and end tags (in this case <title> and </title>) becomes the title.

Next comes the body section, bounded by the <body> and </body> tags. The <p> and </p> tags specify what is displayed in the browser. The document is closed with a </html> tag.

```
<?xml version = "1.0"?>
<!DOCTYPE html PUBLIC "-//W3C//DTD XHTML Basic 1.0//EN"
 "http://www.w3.org/TR/xhtml;-basic/xhtml-basic10.dtd">

<html xmlns = "http://www.w3.org/1999/xhtml">
 <head>
 <title> Welcome</title>
 </head>
 <body>
<!-- displays a welcome message-->

 <p>
 Welcome to XHTML Basic!</p>
 </body>
 </html>
```

Example 10.1

Wireless Markup Language (WML)

Figure 10.1 XHTML Basic

Headers

XHTML Basic provides for six headers or header elements, **h1** through **h6**. The **h1** header is the largest, and **h6** is the smallest. Example 10.2 shows the use of the six headers.

Chapter 10

```
<?xml version = "1.0"?>
<!DOCTYPE html PUBLIC "-//W3C//DTD XHTML Basic 1.0//EN"
 "http://www.w3.org/TR/xhtml;-basic/xhtml-basic10.dtd">

<html xmlns = "http://www.w3.org/1999/xhtml">
 <head>
 <title> Using Headers</title>
 </head>
 <body>

 <h1> Level 1 header</h1>
 <h2> Level 2 header</h2>
 <h3> Level 3 header</h4>
 <h4> Level 4 header</h4>
 <h5> Level 5 header </h5>
 <h6> Level 6 header</h6>


 </body>
 </html>
```

Example 10.2

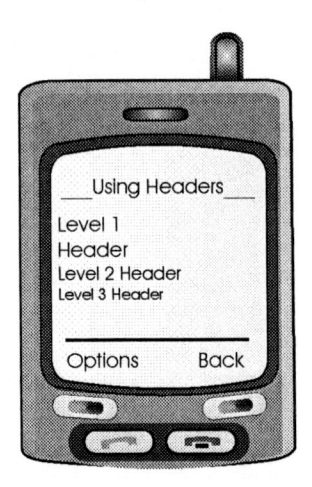

Figure 10.2 Heading Example

Wireless Markup Language (WML)

Linking

The hyperlink is a very important part of XHTML Basic features. The hyperlink provides references, or links, to other resources such as XHTML Basic documents and images. In XHTML Basic, both text and images can be hyperlinks.

The **a** (anchor) element creates links. The text or image between the <a>...</a> tags is the hyperlink that appears on the document. The **href** attribute specifies the location of the linked source, such as a web page or a file. Example 10.3 shows a XHTML Basic program that uses the hyperlink to a web site.

```
<?xml version = "1.0"?>
<!DOCTYPE html PUBLIC "-//W#C//dtd XHTML Basic 1.0//EN"
 "http://www.w3.org/TR/xhtml;-basic/xhtml-basic10.dtd">

<html xmlns = "http://www.w3.org/1999/xhtml">
 <head>
 <title> Hyper Links</title>
 </head>
 <body>

 <p><strong>Select a website</strong><br/>
 <a href="http://www.javasoft.com">
 Get Java Runtime</a><br/>
 <a href="http://www.microsoft.com">
 Go to Microsoft</a><br/>
 <a href="xhtmlex1.html">
 Go to Local Page</a>

 </p>

 </body>
 </html>
```

Example 10.3

Chapter 10

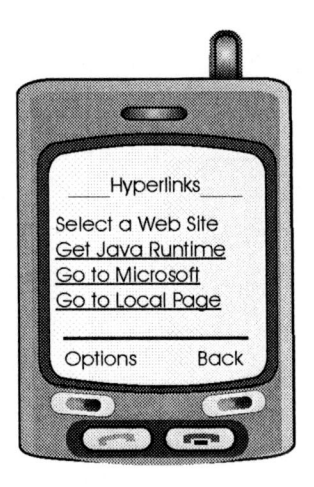

Figure 10.3 Hyperlink Example

In example 10.3 the tags ... are used to make the words within them bold or emphasized. Also, a <br/> command, same as WML and HTML is used to force a new line after each hyperlink. Also, the third hyperlink is a relative link to another xhtml document in the same folder.

Images

The two most popular image formats that Web developers use are GIF and JPEG. In WML documents there also are WBMP formats used for images. In general, wireless web clients cannot display large images due to the small screen size. Therefore, care must be used in selecting the images used for a wireless device.

The **img** element is used to insert an image into a document. The **src** attribute of the **img** element specifies the image file's location. Also, there are optional attributes **width** and **height** that can be used to specify the image's height and width in pixels. If these attributes are omitted, the actual size of the image will be used. There is also another required attribute besides **src**, which is **alt**, which is used by the browser to display the value of the **alt** attribute if it cannot get or display the image. Example 10.4 shows

215

an XHTML Basic document using the img element. Note that this image is in the same directory as the XHTML Basic document, and therefore only the relative address of the image is given.

```
<?xml version = "1.0"?>
<!DOCTYPE html PUBLIC "-//W3C//DTD XHTML Basic 1.0//EN"
 "http://www.w3.org/TR/xhtml-basic/xhtml-basic10.dtd">

<html xmlns = "http://www.w3.org/1999/xhtml">
 <head>
 <title> Using an Image</title>
 </head>
 <body>

 <p>
 <img src = "ppc1.gif" height ="40" width="40"alt="pda image"/>

 </p>

 </body>
 </html>
```

Example 10.4

Chapter 10

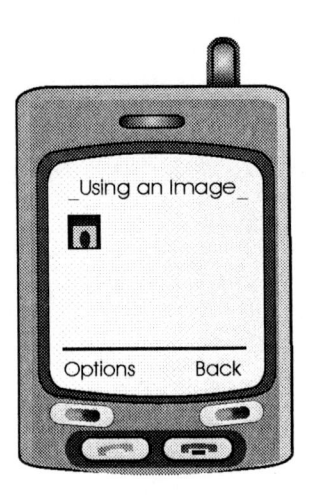

Figure 10.4 Image Example

Like WML, some HTML Basic elements are called empty elements which means they contain attributes only, and do not markup text (there is no text between the start and end tags). Empty elements such as img must be terminated with the / inside the closing right bracket (>) or by explicitly including the end tag.

Images can also be used as hyper links just by inserting the img element within the a tags. See example 10.5.

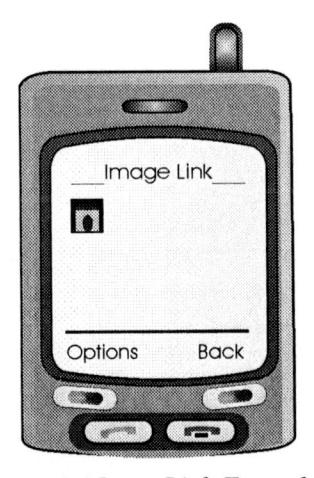

Figure 10.5 ImageLink Example

Wireless Markup Language (WML)

```
<?xml version = "1.0"?>
<!DOCTYPE html PUBLIC "-//W3C//DTD XHTML Basic 1.0//EN"
 "http://www.w3.org/TR/xhtml-basic/xhtml-basic10.dtd">
<html xmlns = "http://www.w3.org/1999/xhtml">
 <head>
 <title> Image Link</title>
 </head>
 <body>
 <p>
<a href="http://www.mindspring.com">
 <img src = "mshead.gif" height ="40" width="40"alt="Mindspring image
 </a>
 </p>
 </body>
 </html>
```

Example 10.5

Special Characters

When marking up text, certain characters or symbols (e.g. >) may be hard to embed directly into an XHTML document. Some keyboards may not provide these symbols or there presence may cause syntax errors. For example, the line

<p> if x < 9 then increment x by 1</p>

causes a syntax error because it uses the < character which is reserved for start and end tags. XHTML Basic provides special characters or entity references (in the form &code;) for representing these characters. The previous line corrected is:

<p> if x < 9 then increment x by 1 </p>

in which the special character < represents the < symbol. For a complete list of character entity references, see the web site :
www.w3.org/TR/REC-html40/sgml/entities.html .

Chapter 10

Example 10.6 shows the use of some special characters used in XHTML Basic.

```
<?xml version = "1.0"?>
<!DOCTYPE html PUBLIC "-//W3C//DTD XHTML Basic 1.0//EN"
 "http://www.w3.org/TR/xhtml-basic/xhtml-basic10.dtd">
<html xmlns = "http://www.w3.org/1999/xhtml">
 <head>
 <title>Special Characters</title>
 </head>
 <body>
 <p>
 Statistically &gt; &frac12; of the
people on the web are &lt; 50.
</p>
 </body>
 </html>
```

Example 10.6

Figure 10.6 Special Character Example

219

Tables

The **table** element defines a table. XHTML Basic tables lack the advanced features of XHTML like borders and nesting. Also, some microbrowsers render XHTML Basic tables without alignment. You can add the special character ** ** to add spaces in a column.

The attribute **summary** of the **table** element describes the table's contents. Speech devices use this attribute to make the table more accessible to users with visual impairments.

The **caption** element describes the table's content and helps text-based browsers interpret the table data. Most browsers put the text inside the <caption> tags above the table. Attribute summary and element caption are two of many XHTML Basic features that make Web pages more accessible to users with disabilities.

The Pixo Microbrowser does not support the table header element , **th**. The **th** element causes other browsers to render the text within the header cells in bold. You can also use the **strong** element to create the same effect.

Each **tr** element defines an individual table row. Data cells contain individual pieces of data and they are defined with **td** (table data) elements.

Example 10.7 shows a table in an XHTML Basic document.

Chapter 10

```
<?xml version = "1.0"?>
<!DOCTYPE html PUBLIC "-//W3C//DTD XHTML Basic 1.0//EN"
 "http://www.w3.org/TR/xhtml-basic/xhtml-basic10.dtd">
<html xmlns = "http://www.w3.org/1999/xhtml">
 <head>
 <title>Basic Table</title>
</head>
<body>
 <table summary="This table provides information about the price of fru
<caption> Price of Fruit</caption>

<tr>
 <td>Apple</td>
 <td>$0.25</td>
 </tr>
<tr>
 <td>Orange</td>
 <td>$0.50</td>
 </tr>
<tr>
 <td>Banana</td>
 <td>$1.00</td>
</tr>
</table>

 </body>
</html>
```

Example 10.7

Wireless Markup Language (WML)

Figure 10.7 Table Example

Unordered Lists

Unordered lists can be created in XHTML Basic by using the **ul** element. Unordered lists are lists that does not order its items by letter or number. Each entry in an unordered list is an **li** element. Most wireless clients show these elements with a line break and a bullet mark at the beginning of the line. Example 10.8 shows an XHTML Basic document using an unordered list.

Chapter 10

```
<?xml version = "1.0"?>
<!DOCTYPE html PUBLIC "-//W3C//DTD XHTML Basic 1.0//EN"
 "http://www.w3.org/TR/xhtml-basic/xhtml-basic10.dtd">
<html xmlns = "http://www.w3.org/1999/xhtml">
 <head>
 <title>Unordered List</title>
 </head>
<body>
 <p>Select a Web site</p>

 <ul>
 <li>
 <a href="http://www.prenhall.com">
 Prentice Hall
 </a>
 </li>
 <li>
 <a href="http://www.microsoft.com">
 Microsoft
 </a>
 </li>
 </ul>
 </body>
 </html>
```

Example 10.8

Figure 10.8 Unordered List Example

Nested and Ordered Lists

An ordered list also exists in XHTML Basic, using the **ol** element. An ordered list is enumerated or numbered by the browser. It also uses the **li** element for items in the list, as does the unordered list.

Lists can be nested in XHTML Basic in order to show a hierarchical relationship between items. Both unordered lists and ordered lists can be nested. A nested list is contained within a list element of another list. Most web browsers show nested lists by indenting the list one level.

Example 10.9 shows an ordered list nested within an unordered list.

Chapter 10

```
<?xml version = "1.0"?>
<!DOCTYPE html PUBLIC "-//W3C//DTD XHTML Basic 1.0//EN"
 "http://www.w3.org/TR/xhtml-basic/xhtml-basic10.dtd">
<html xmlns = "http://www.w3.org/1999/xhtml">
 <head>
 <title>Nested Lists</title>
 </head>
<body>
 <p>Select a Web site</p>

 <ul>
 <li>
 <a href="http://www.prenhall.com">
 Prentice Hall
 </a>
 <ol>
 <li> Text Books</li>
 <li> Training</li>
 </ol>
 </li>
 <li>
 <a href="http://www.microsoft.com">
 Microsoft
 </a>
 </li>
 </ul>
 </body>
 </html>
```

Example 10.9

Wireless Markup Language (WML)

Figure 10.9 Nested List Example

Simple XHTML Basic Forms

When browsing web sites, the user is often asked to provide information such as email address, search keywords, and addresses. XHTML Basic provides a mechanism, called forms, for collecting such information. These forms send the data to the web server, which passes it to a CGI script, usually written in C, Perl, or one of several other languages. The data is sent to the CGI script when the submit button is pushed on the XHTML Basic document. The script then processes the data and returns information in the form of an XHTML Basic document to the web browser.

Example 10.10 shows a XHTML Basic document that asks for the users name and email address.

```xhtml
<?xml version = "1.0"?>
<!DOCTYPE html PUBLIC "-//W3C//DTD XHTML Basic 1.0//EN"
 "http://www.w3.org/TR/xhtml-basic/xhtml-basic10.dtd">
<html xmlns = "http://www.w3.org/1999/xhtml">
 <head>
 <title>Feedback Form</title>
 </head>
 <body>
 <h3>Feedback Form</h3>

 <form method = "post" action="/cgi-bin/nameemail">

<p>
 <label>
 Name:
 <input type="text" name="name"/><br/>
 </label>
 <label>
 email:
 <input type="text" name="email"/><br/>
 </label>

 <input type="submit" value="Submit"/>
 <input type="reset" value="Clear"/>
 </p>
 </form>

  </body>
 </html>
```

Example 10.10

Wireless Markup Language (WML)

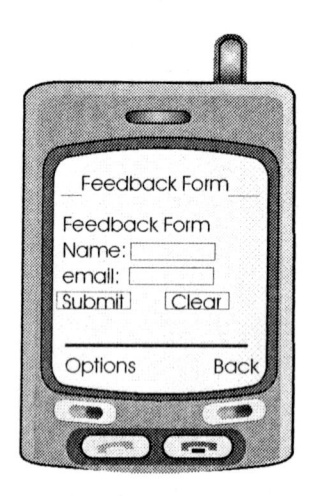

Figure 10.10 Forms Example

Forms can contain visual and non-visual components. Visible components include clickable buttons and other graphical user interface components for the user to interact. Non-visual components, called hidden inputs, store data that the document author wishes to pass to the cgi script.

In the **form** element, the **method** attribute specifies how the form's data is sent to the web server. Using method = "post" appends form data to the browser request, which contains the protocol (i.e.HTTP) and the resources URL. Scripts that reside on the web server can access the form data sent as part of the request. The other possible value for **method** is "get" which appends the form data directly to the end of the URL. The **action** attribute of the **form** tag specifies the URL of a script on the web server.

The **input** element specifies data to be provided to the script. The input element can have **type** = "hidden", which allows the document author to send form data that is not entered by the user to a script. The input attribute **name** specifies the variable name for this data, and the attribute **value**, specifies a value for that variable inputted.

Another input **type** is "text". This inserts a text box into the document. Users can type data into these text boxes. The **label** element displays information about the input element's purpose next to the text box.

Chapter 10

The input **types** of "submit" and "reset", create buttons on the display that cause the data to be submitted (to the cgi script) or cleared out of the forms. The **value** attribute specifies the text for the buttons.

More Complex XHTML Basic Forms

XHTML Basic also allows for the use of more complex forms for inputting data from a user. One is **textarea** that allows more text to be entered than with text boxes. Also, checkboxes and radio buttons offer pre-set options to the user.

The **textarea** element inserts a multiline text box, called a text area into the form. The **rows** attribute specifies the number of rows, and the **cols** attribute specifies the number of columns. To display a default text in the text area, place the text between the <textarea>...</textarea> tags. Default text can be specified in other input types, such as text boxes, by using the **value** attribute.

The input type **checkbox** enables users to select items from a set of options. When a user selects a checkbox, a check mark appears in the check box. Checkbox inputs can be used individually or in groups. When used in groups, each checkbox input is given the same **name**.

The input type **radio** is similar to checkboxes, except that only one radio button in a group of radio buttons can be selected at any time. All radio buttons in a group have the same **name** and have different **value** attributes.

Another input form is the **select** element. The **select** element provides a drop down list from which a user can select an item. The **name** attribute identifies the drop-down list. The **option** element adds items to the drop-down list. The **option** element's **selected** attribute specifies which item is the initially selected item in the **select** element.

Example 10.11 shows the use of an input type of textarea, as well as an input type of checkbox.

Wireless Markup Language (WML)

```
<?xml version = "1.0"?>
<!DOCTYPE html PUBLIC "-//W3C//DTD XHTML Basic 1.0//EN"
 "http://www.w3.org/TR/xhtml-basic/xhtml-basic10.dtd">
<html xmlns = "http://www.w3.org/1999/xhtml">
 <head>
 <title>More Complex Forms</title>
</head>
<body>

<form method = "post" action="/cgi-bin/geninput">

<p>
Food you like:<br/>
<label>
Fruit
<input type="checkbox " name="foodliked" value="fruit"/><br/>
</label>
<label>
Meat
<input type="checkbox" name="foodliked" value="meat"/><br/>
</label>
<label>
Cheese
<input type="checkbox" name="foodliked" value="cheese"/><br/>
</label>

Comments:
<textarea name="comments" rows="4" cols="15">
</textarea>
<input type="submit" value="Submit"/>
<input type="reset" value="Clear"/>
</p>
</form>

</body>
</html>
```

Example 10.11

Chapter 10

Figure 10.11 Textarea Example

Example 10.12 shows the use of the input type **radio**, as well as the **select** element.

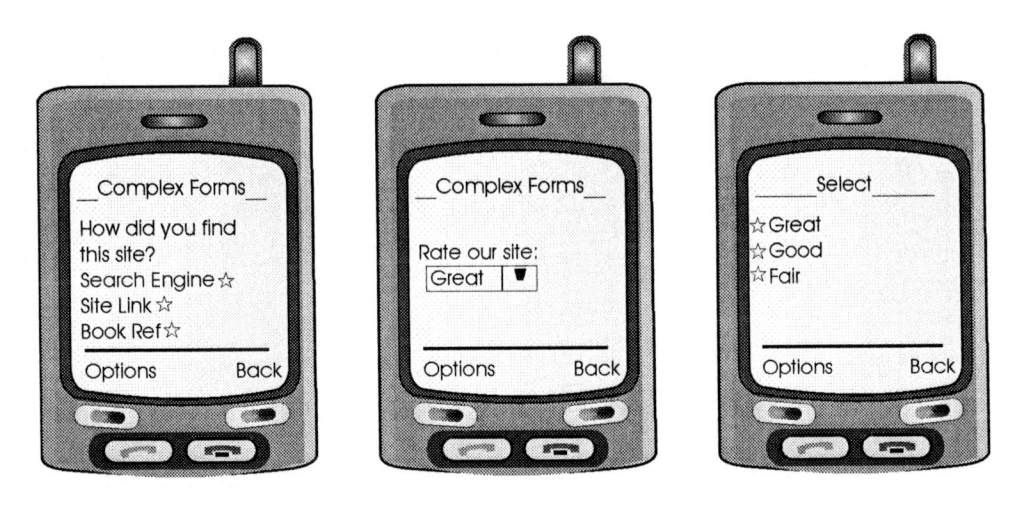

Figure 10.12 Radio and Select Example

231

Wireless Markup Language (WML)

```
<?xml version = "1.0"?>
<!DOCTYPE html PUBLIC "-//W3C//DTD XHTML Basic 1.0//EN"
 "http://www.w3.org/TR/xhtml-basic/xhtml-basic10.dtd">
<html xmlns = "http://www.w3.org/1999/xhtml">
 <head>
 <title>Complex Forms</title>
 </head>
 <body>

 <form method = "post" action="/cgi-bin/geninput">

 <p>
 How did you find this site?<br/>
 <label>
 Search Engine
 <input type="radio" name="howtosite" value="search engine"/><br/>
 </label>
 <label>
 Site Link
 <input type="radio" name="howtosite" value="link"/><br/>
 </label>
 <label>
 Book Ref
 <input type="radio" name="howtosite" value="book"/><br/>
 </label>
 <label>
 Rate our site:
 <select name="rating">
 <option selected ="selected">Great</option>
 <option>Good</option>
 <option>Fair</option>
 </select>
 </label>
 </p>
 </form>

 </body>
 </html>
```

Example 10.12

Chapter 10

cHTML

The wireless programming language cHTML is a subset of HTML 2.0, 3.2 and 4.0. It is used in wireless devices that are on the i-mode network, served by the internet service provider DoCoMo. cHTML supports fewer character fonts and styles than HTML, and it also does **not** support: image maps, tables, cookies, frames, or JPEGS. cHTML also does not use style sheets or scripts. Like XHTML Basic, it must use cgi-bin scripts written in Perl, C++, or Java for processing input data or doing other processing of data.

cHTML documents are created with a text editor such as Wordpad or Notepad, but are also included in the newer SDKs such as the Nokia Toolkit 4.0. cHTML documents usually use the .html or .htm extension for its files.

When programming for wireless devices, as with other wireless device languages, you must take into consideration the phone's small display sizes, memory and bandwidth.

The cHTML document starts with a DOCTYPE element that indicates that the following document conforms to the latest specifications for cHTML. Comments are the same in cHTML as both XHTML Basic and WML and begin with <!— and end with —>.

The cHTML document uses the **<html>**...**</html>** tags to surround the document. Every cHTML document uses a **head** element to specify information about the document. It also has a **body** element that contains the document content. The <title> tags are used to specify a title for the document that usually appears at the top of the screen.

The <p>..</p> paragraph elements enable you to place text on the display.

Headers

A header is a simple form of text formatting that varies the size of the text with the header's level. There are six header elements (**h1** to **h6**) that can be used.

Wireless Markup Language (WML)

The actual size of the text for each header level can vary significantly with the browser and the device. Example 10.13 shows a basic cHTML document using a few of the header elements.

```
<!DOCTYPE html PUBLIC "-//W3C//DTD Compact HTML 1.0 Draft//EN">

<html>
 <head>
 <title>Welcome</title>
 </head>
 <body>

 <h1>Level 1 </h1>
 <h2>Level 2</h2>
 <h3>Level 3</h3>

 </body>
</html>
```

Example 10.13

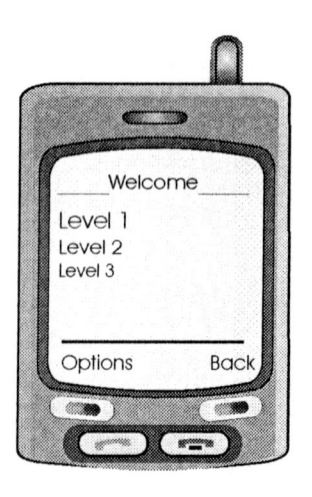

Figure 10.13 Header Example

Linking

Linking to other web pages is done using the <a>...</a> anchor element. A required attribute for the **a** element is the href attribute. This specifies the address that the link goes to. Example 10.14 shows the use of the anchor element in a cHTML document.

```
<!DOCTYPE html PUBLIC "-//W3C//DTD Compact HTML 1.0 Draft//EN">

<html>
 <head>
 <title>Computer Sites</title>
 </head>
 <body>

 <p>
 Pick a computer helpful site<br/>

 <a href="http://www.cnet.com">CNET</a><br/>
 <a href="http://www.tigerdirect.com">TigerDirect</a><br/>
 <a href="http://www.drivers.com">Drivers</a>
 </p>
 </body>
</html>
```

Example 10.14

Wireless Markup Language (WML)

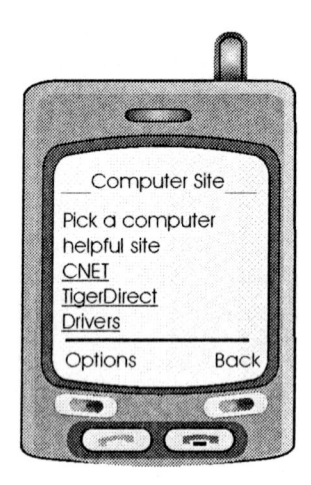

Figure 10.14 Link Example

Images

I-mode handsets only support gif images. The latest I-mode phones support interlaced gif, ainimated gif, and transparent gif. Many I-mode handsets also support color.

The **img** element specifies the location of the image file. This is done with the **src** attribute (a required attribute). The **height** and **width** attributes can be used to specify the size of an image in pixels.

Example 10.15 shows the use of an image in a cHTML document.

Chapter 10

```
<!DOCTYPE html PUBLIC "-//W3C//DTD Compact HTML 1.0 Draft//EN">

<html>
 <head>
 <title>Wireless</title>
 </head>
 <body>

 <p>
 <img src="mshead.gif" height="40" width="40"
 alt="Mindspring Logo"/>
 </p>
 </body>
</html>
```

Example 10.15

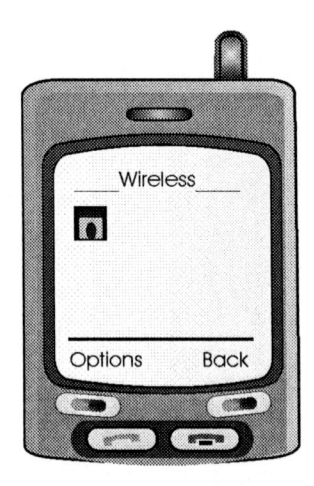

Figure 10.15 Image Example

Wireless Markup Language (WML)

Changing Text Color

Text color in cHTML can be changed with the **font** element. The attribute **color** in the **font** element sets the color for the text. The colors available are the same as the standard HTML colors. cHTML can recognize the standard names such as "red" or "blue" or by the hexidecimal value such as #0000FF for blue.

Also the center element is used in cHTML to center the text in the display. Normally text is left adjusted by default.

Example 10.16 shows the use of the center element as well as font color.

```
<!DOCTYPE html PUBLIC "-//W3C//DTD Compact HTML 1.0 Draft//EN">
<html>
 <head>
 <title>Wireless</title>
 </head>
<body>
 <h1><center>Wireless</center></h1>
 <p>
 <font color="red">cHTML supports text formatting</font>
 <font color="blue">and colors</font>
 <font color="green">as well</font>
 </p>
 </body>
</html>
```

Example 10.16

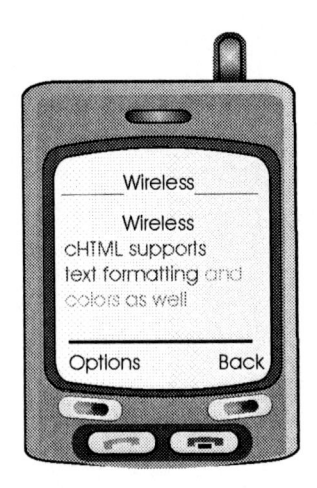

Figure 10.16 Font Color Example

Special Characters and More Line Breaks

The normal line break element **br** is available in cHTML. Also, the horizontal rule element **hr** is usable in cHTML and it inserts a horizontal line across the display.

As for special characters, they can be inserted in the code in cHTML documents by using the form **&code;**. For example & insets the special character ampersand and **©** inserts the copyright symbol. These special characters can also be inserted using their hexidecimal code as well. For example, **&** inserts an ampersand since #38 is the hexidecimal code for an ampersand.

Example 10.17 shows the use of the **hr** element as well as certain special characters.

Wireless Markup Language (WML)

```
<!DOCTYPE html PUBLIC "-//W3C//DTD Compact HTML 1.0 Draft//EN">

<html>
 <head>
 <title>Spec. Char</title>
 </head>
 <body>

 <p>
 Information on this site is &copy; ABC &amp; Co.<br/>
 <hr/>

There are &gt; 24M i-mode users.
 </p>
 </body>
 </html>
```

Example 10.17

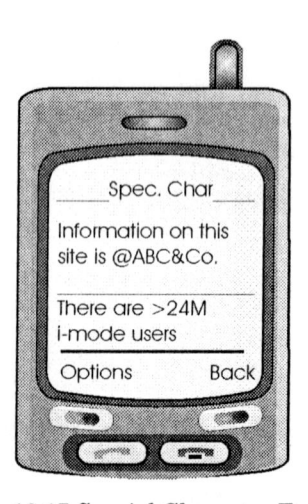

Figure 10.17 Special Character Example

Chapter 10

Unordered Lists

Like XHTML Basic, cHTML allows the use of unordered lists with the **ul** element. And, the **li** element is used to specify each entry in the unordered list. Each entry in an unordered list usually is preceded with a bullet in most browsers, and followed by a line break.

Example 10.18 shows the use of the unordered list capability.

```
<!DOCTYPE html PUBLIC "-//W3C//DTD Compact HTML 1.0 Draft//EN">

<html>
 <head>
 <title>Unordered List</title>
 </head>
 <body>

 <p>
 Search Engines:
 <ul>
 <li><a href="http://www.yahoo.com">Yahoo</a></li>
 <li><a href="http://www.google.com">Google</a></li>
 <li><a href="http://www.lycos.com">Lycos</a></li>
 </ul>

 </p>
 </body>
 </html>
```

Example 10.18

Wireless Markup Language (WML)

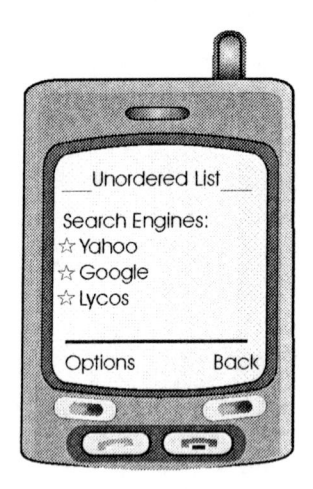

Figure 10.18 Unordered List Example

Nested and Ordered Lists

Lists can be nested, in order to show a hierarchical form or outline form, both unordered lists as well as ordered lists. Ordered lists are specified by the **ol** element, with the **li** element used to specify the items in the list. By default, most browsers number the items in an ordered list :1,2,3... In order to change this, the attribute **type** can be used within an **li** element, to change the numbering for that item. If the attribute **type** is set to "I", the sequence used is I,II,III... If **type** is set to "i" the sequence is i,ii,iii... Also, A and a can be used for the value of **type** to get an alphabetical list in upper or lower case.

Example 10.19 shows nested and ordered lists.

```html
<!DOCTYPE html PUBLIC "-//W3C//DTD Compact HTML 1.0 Draft//EN">

<html>
 <head>
 <title>Nested Lists</title>
 </head>
 <body>

 <p>
 Best Internet Features
 <ul>
 <li>Meet People</li>
 <li>New Media</li>
 <ul>
 <li>New games</li>
 <li>New Apps</li>
 <ol>
 <li>For Business</li>
 <li>For pleasure</li>
 </ol>

 </ul>
 </ul>

 </p>
 </body>
 </html>
```

Example 10.19

Wireless Markup Language (WML)

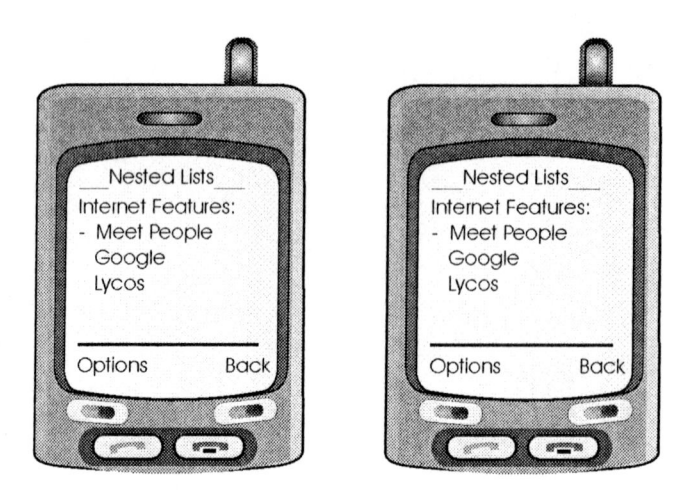

Figure 10.19 Nested Lists Example

cHTML Forms

cHTML uses the same forms for inputting data from a user that XHTML Basic does, presented earlier in this chapter. The **input** element can have the same values for its **type** attribute as XHTML Basic can. These are: text, password, checkbox, radio, hidden, submit, and reset. They perform exactly like XHTML Basic, and, of course, HTML.

The **select** element also exists for cHTML as it does in XHTML Basic. It creates a drop down list, and the **option** element is used to create the choices in the list.

The **textarea** element also is available in cHTML to use to create a larger input area than the **input** element with type=text. This is also the same as XHTML Basic.

Examples of all of these forms can be seen earlier in this chapter under the XHTML Basic section.

Chapter 10

Summary

- Unlike WML which has WMLScript, XHTML Basic and cHTML rely on cgi-bin programs written in perl, java, or C++ to process data inputted from a user.

- XHTML Basic is derived from XML, and is therefore similar to WML.

- XHTML Basic documents are created with a text editor, but are now included in some of the newer SDK's for wireless devices.

- XHTML Basic documents use start and end tags, like HTML and WML.

- The head section contains information about the XHTML Basic document.

- The body section contains the content that a browser displays.

- Comments in XHTML Basic are bounded by the tags <!— and —>.

- XHTML Basic provides for six headers or header elements, **h1** through **h6**.

- The **a** element creates a hyperlink for accessing other resources on the internet. The **href** attribute specifies the location of the linked source.

- XHTML Basic displays images using the **img** element. The attribute **src** specifies the location of the image. XHTML Basic handles gif and jpeg formats.

- Special characters are handled in XHTML Basic in the form &code;. The code can be an abbreviation for the special character, or a hex value for the special character.

Wireless Markup Language (WML)

- XHTML Basic allows for tables, but tables don't have borders and cannot be nested. It uses the **table** element, and **tr** and **td** elements specify row and cell, respectively.

- XHTML Basic allows for unordered lists, using the **ul** element and the **li** element. Unordered lists are not ordered by number or letter.

- Ordered lists also exist in XHTML Basic, using the **ol** element and the **li** element. Both ordered and unordered lists can be nested in XHTML Basic.

- XHTML Basic uses forms for inputting data from the user to cgi scripts residing on the web server. The **input** element specifies data to be provided to the script. The attribute **name** specifies the variable name for the data. The **type** attribute can be: text, hidden, submit, reset, checkbox, and radio.

- XHTML Basic also uses the **textarea** element for inputting multi line text.

- Another element in XHTML Basic is the **select** element which creates a drop down list.

- cHTML is a subset of HTML. cHTML supports fewer character fonts, and styles than HTML, and it also does not support image maps, tables, cookies, frames, or jpegs.

- cHTML is very similar to XHTML Basic, and they share many of the same elements and features.

- cHTML provides six header elements, **h1 – h6**.

- cHTML uses the <a>...</a> **anchor** element for linking to other web pages.

Chapter 10

- cHTML only supports gif images using the **img** element.

- Text color can be changed in cHTML using the **color** attribute in the **font** element.

- cHTML supports the **br** element and the **hr** element for line break and for a horizontal rule.

- Special characters can be used in cHTML just like in XHTML Basic using the form **&code;** .

- cHTML, like XHTML Basic, allows for the use of unordered lists with the **ul** element and the **li** element.

- cHTML also allows for ordered lists using the **ol** element and the **li** element. Also, nesting is allowed of both ordered and unordered lists.

- cHTML uses the same forms for inputting user data that XHTML Basic uses.

Wireless Markup Language (WML)

Questions

1.) Of the three main languages used in programming wireless devices, which two are being merged in WAP 2.0 specifications?

2.) How does XHTML Basic and cHTML process data inputted by the user?

3.) What basic features are not included in XHTML Basic that are in XHTML?

4.) What is wrong with the XHTML Basic code below?

   <WML_MISC_10-1>

5.) How many headers are allowed in XHTML Basic, and what are the elements for them?

6.) Write a correct line in XHTML Basic that links to the web site www.msn.com.

7.) What attributes are used in the **img** element that can size the image?

8.) Write the XHTML Basic line of code that prints: if x > y then z < 3.

9.) What two features are not included in the table element in XHTML Basic?

10.) What two elements are used in tables in XHTML Basic?

11.) How do unordered lists differ from ordered lists?

12.) Can ordered lists be nested in XHTML Basic?

13.) List the allowable **type** attribute values for the input element in XHTML Basic.

Chapter 10

14.) What element is used if a larger text input area is needed in XHTML Basic?

15.) What features does cHTML not support?

16.) What type of images does cHTML support?

17.) Write a line of code for cHTML that changes the text color to red.

18.) What element is used to put a horizontal line across the display?

19.) What element is used in both ordered lists and unordered lists in cHTML?

Wireless Markup Language (WML)

Problems

1.) Write a XHTML Basic document that includes comments and displays a greeting to the user.

2.) Write a XHTML Basic program that links to three different web sites.

3.) Write a XHTML Basic program that shows an image that is sized using the **width** and **height** attributes, and is also a link to a web site : www.drivers.com.

4.) Write a XHTML Basic program that displays the line: ½ is > ¼.

5.) Write a XHTML Basic program that has a table with 3 rows and 2 columns with the words one, two, three, four, five and six in the cells.

6.) Write a XHTML Basic program that has an unordered list specifying three top news magazines.

7.) Write a XHTML Basic program that has a nested ordered list within a nested ordered list where the main list is of three car manufacturers, and the inner list has three models of one of the manufacturers.

8.) Write a XHTML Basic program that requests the user to input his name and phone number using type = text, and also passes a hidden code (abc123) to the script.

9.) Use the textarea element in a XHTML Basic program to input a description of a web site.

10.) Generate a drop down list in a XHTML Basic program that specifies a list of choices in types of ice cream. (use 3 types of ice cream)

11.) Repeat problem 1 in cHTML.

12.) Repeat problem 2 in cHTML.

Chapter 10

13.) Repeat problem 3 in cHTML.

14.) Repeat problem 4 in cHTML.

15.) Repeat problem 5 in cHTML.

16.) Repeat problem 6 in cHTML.

17.) Repeat problem 7 in cHTML.

18.) Repeat problem 8 in cHTML.

19.) Repeat problem 9 in cHTML.

20.) Repeat problem 10 in cHTML.

Wireless Markup Language (WML)

Appendix I

WML Commands and Attributes

Command	Attributes
<a>...</a>	**href** title
<access/>	domain path
<anchor>...</anchor>	title
...	
<big>...</big>	
<card>...</card>	id class title newcontext ordered onenterforward onenterbackward ontimer

Wireless Markup Language (WML)

<do>...</do>   **type**
                                label
                                name
                                optional

<em>...</em>

<fieldset>...</fieldset>   title

<go>...</go>   **href**
 (<go/>)   sendreferer
                                method
                                accept-charset

<head>...</head>

<i>...</i>

<img/>   **alt**
                                src
                                localsrc
                                align
                                height
                                width
                                vspace
                                hspace

<input/>   **name**
                                title
                                type
                                value
                                default
                                format
                                emptyok
                                size
                                maxlength
                                tabindex

Appendix I

<meta/>	**name | http-equiv | user-agent** **content** scheme forua
<noop/>	
<onevent>…</onevent>	**type**
<optgroup> </optgroup>	title
<option>…</option>	title value onpick
<p>…</p>	align mode
<postfield/>	**name** **value**
<prev>…</prev> (<prev/>)	
<refresh>…</refresh>	
<select>…</select>	title multiple name value iname ivalue tabindex
<setvar/>	**name** **value**

Wireless Markup Language (WML)

<small>...</small>

<strong>...</strong>

<table>...</table>   align
title
columns

<td>...</td>

<template>...</template>   onenterforward
onenterbackward
ontimer

<timer/>   name
value

<tr>...</tr>

<u>...</u>

<wml>...</wml>   xml:lang

Appendix II

WML Libraries

WMLScript is a client-side scripting language. For reference, this appendix lists all functions supported by WMLScript. Please note the following:

Like WML, WMLScript is case-sensitive.

WMLScript functions are members of function libraries, and function names are always specified as library.function (for example, String.length()).

All functions and libraries listed here are supported on all devices except where noted.

Strings may be enclosed within double quotes (as in "string") or single quotes (as in 'string').

Lang Library

The lang library contains functions closely related to the WMLScript language core.

Wireless Markup Language (WML)

abort

Description: This function will interrupt the processing of the WMLScript and return the value of errorDescription (which must be a string) to the device.

Syntax: Lang.abort(errorDesciption)

Example: The following example interrupts WMLScript processing, and displays the error message:

```
// Compare passwords
if (password1!=password2) {
// Passwords do not match
 Lang.abort("Passwords do not match!");
}
```

abs

Description: This function returns the absolute value of a specified number. If value is of type floating point, then the result is of type floating point. If value is not one of these types, then "invalid" is returned.

Syntax: Lang.abs(value)

Example: The following example returns the absolute value of the value of a specified variable:

```
// Get absolute value of a
var b=Lang.abs(a);
```

Appendix II

characterSet

Description: The Internet Assigned Numbers Authority (IANA) has assigned integer values for all character sets. CharacterSet returns this value.

Syntax: Lang.characterSet()

Example: The following example checks and returns the character set supported by the WMLScript interpreter:

```
//Check character set
function charSetTest()
 if (Lang.characterSet()==4)[
 return ("Western Europe!");
} else {
 return ("unkown");
}
```

exit

Description: This will end the rendering of the WMLScript code and present the string value. You can use this function to normally exit your WMLS code where needed.

Syntax: Lang.exit(value)

Example: The following example exits the current WMLScript and returns the specified string to the device:

```
// Check if okay to proceed, exit if not
if (!proceed){
 Lang.exit("Cannot proceed as requested");
}
```

Wireless Markup Language (WML)

float

Description: This function returns either true or false. If floating points are supported, the result is true. Otherwise, the result is false.

Syntax: Lang.float()

Example: The following example returns true if floating points are supported by the WAP browser:

```
// Check if floating point is supported before continuing
if (!Lang.float()){
 Lang.abort("Cannot perform calculation");
}
```

isFloat

Description: If value can be converted into a floating point, this function returns true. Otherwise, the result is false.

Syntax: Lang.isFloat(value)

Example: The following converts a floating point number to its closest integer (if it is a floating number):

```
// Is "a" a float?
if (Lang.isFloat(a)){
 // yes it is, convert to integer
 var b=Float.round(a);
}
```

Appendix II

isInt

Description: If value can be converted into a integer, this function returns true. Otherwise, the result is false.

Syntax: Lang.isInt(value)

Example: The following example checks to see that a specified value is an integer (and not a float); if it is not it prompts for the value again.

```
// Is "a" an int?
if (!Lang.isInt(a)){
 //prompt for it again
 a=Dialogs.prompt("Please enter a valid integer value", "");
}
```

max

Description: Evaluates two values and determines which is the larger number. Whether the result is an integer or floating point is determined by the type of value1 and value2.

Syntax: Lang.max(value1,value2)

Example: The following example saves the greater of two specified values:

```
// Get greater value
var a = Lang.max(var1,var2);
```

Wireless Markup Language (WML)

maxInt

Description: This will return the largest supported integer value.

Syntax: Lang.maxInt()

Example: The following example initializes an integer value to the highest support value:

```
//Initialize to greatest possible value
var a = Lang.maxInt();
```

min

Description: Evaluates two values and determines which is the smaller number. Whether the result is an integer or floating point is determined by the type of value1 and value2.

Syntax: Lang.min(value1,value2)

Example: The following example saves the lesser of two specified values:

```
// Get lesser value
var a = Lang.min(var1,var2);
```

minInt

Description: This will return the smallest supported integer.

Syntax: Lang.minInt()

Example: The following example initializes an integer value to the smallest support value:

```
// Initialize to smallest possible value
var a = Lang.minInt();
```

Appendix II

parseFloat

Description: Converts a string to type floating point. Parsing ends at the first character that cannot be parsed.

Syntax: Lang.parseFloat()

Example: The following example returns 2.500000e+000:

// Prompt for temperature and parse into a float
var temp=Lang.parseFloat(Dialogs.prompt("Temperature:",""));

parseInt

Description: Converts a string to type integer.

Syntax: Lang.parseInt(value)

Example: The following returns the house number portion of an address:

//get house number
var house_num=Lang.parseInt(address1);

random

Description: Generates a random integer greater or equal to 0 and less than or equal to a specified value.

Syntax: Lang.random(value)

Example: The following example is a function that will return a random number between 0 and 100:

// Return a random number between 0 and 100
function GetRandom(){
 return Lang.random(100;
}

Wireless Markup Language (WML)

seed

Description: This initializes a random numeric sequence and also returns an empty string. If value is of type floating point, then Float.int() should be used first to convert the value to an integer.

Syntax: Lang.seed(value)

Example: The following will initialize the random numeric sequence based on the number 42:

```
//Initialize random sequence
Lang.seed(42);
```

Float Library

The Float library contains functions used to manipulate floating point numbers.

ceil

Description: This function returns the smallest integer not less than the specified value.

Syntax: Float.ceil(value)

Example: The following example converts a floating point number to the smallest integer value not less than itself:

```
// Convert to integer
var b=Float.ceil(a);
```

Appendix II

floor

Description: This function returns the largest integer not greater than the specified value.

Syntax: Float.floor(value)

Example: The following example converts a floating point number to the largest integer value not greater than itself:

```
// Convert to integer
var b = Float.floor(a);
```

int

Description: Returns the integer portion of floating point value.

Syntax: Float.int(value)

Example: The following example returns the integer portion of a user-specified value:

```
// Extract integer portion
var b=Float.int(a);
```

maxFloat

Description: This will return the largest supported floating point.

Syntax: Float.maxFloat()

Example: The following example initializes a floating-point value to the highest supported value:

```
//Initialize to greatest possible value
var a=Float.maxFloat();
```

minFloat

Description: This will return the smallest supported floating point.

Syntax: Float.minFloat()

Example: The following example initializes a floating point value to the smallest supported value:

```
// Initialize to smallest possible value
var a = Float.minFloat();
```

pow

Description: Returns the value of a number raised to a specified power.

Syntax: Float.pow(value1,value2)

Example: The following example determines the cubed value of a specified number:

```
// Determine the cubed value of a
var b=Float.pow(a,3);
```

round

Description: Returns the integer that is closest to a specified value. If the two possible results are equidistant from the specified value, the result is the larger of the two.

Syntax: Float.round(value)

Example: The following example rounds a user-provided temperature to the nearest whole number:

```
// Round to nearest whole number
var temp2=Float.round(temp1);
```

sqrt

Description: Returns the square root of value.

Syntax: Float.sqrt(value)

Example: The following example determines the square root of a user-supplied number:

```
// Get square root
var s=Float.sqrt(a);
```

String Library

The String library contains a set of string functions. In WMLScript, a string is essentially an array of characters, the first element of which is at offset 0.

charAt

Description: Returns the character at the specified position in a string.

Syntax: String.charAt(string, index)

Example: The following example returns the first letter of a user's name:

```
// get initial
var initial=String.charAt(first_name,0);
```

Wireless Markup Language (WML)

compare

Description: Compares two strings. Returns 1 if first string is greater than the second string, 0 if both strings are the same, or −1 if first string is less than the first string.

Syntax: String.compare(string1,string2)

Example: The following example (part of a sorting routine) compares two strings and switches them if the first is greater than the second:

```
//Is a greater than b?
if (String.compare(a,b)) {
 //yes, switch a and b
 var x=a;
 a=b;
 b=x;
}
```

elementAt

Description: Returns the element found at a specified location within a string (using a specified delimiter).

Syntax: String.elementAt(string,index,seperator)

Example: The following example returns the first word in a sentence:

```
// get first word
var first_word=String.elementAt(sentence,0," ");
```

Appendix II

elements

Description: Returns the total number of elements in a specified string.

Syntax: String.elements(string, seperator)

Example: The following example determines the number of words in a sentence:

```
// How many words in the sentence?
var num_words=String.elements(sentence," ");
```

find

Description: Determines if a block of text is contained within a specified string. If so, the result is the index of the first character of the first occurrence of the string. If no occurrence is fouind the result is −1.

Syntax: String.find(string,substring)

Example: The following example finds the start of a page's title text:

```
// Find start of page title
var a=String.find(page,"<TITLE>");
```

format

Description: Converts a specified value to a string by using the given formatting provided as a format string. The format string can contain only one format specifier, which can be located anywhere inside the string. If more than one is specified, only the first one (leftmost) is used and the remaining specifiers are replaced by an empty string. The format specifier has the following form:

% [width] [.precision] type

The width argument is a non-negative decimal integer controlling the minimum number of characters printed. If the number of characters in the output value is less than the specified width, blanks are added to the left until the minimum width is reached. The width argument never causes the value to be truncated. If the number of characters in the output value is greater than the specified width, or if width is not given, all characters of the value are printed (subject to the precision argument).

The precision argument specifies a non-negative decimal integer, preceded by a period (.), which can be used to set the precision of the output value. The interpretation of this value depends on the given type:

- d specifies the minimum number of digits to be printed. If the number of digits in the value is less than precision, the output value is padded on the left with zeros. The value is not truncated when the number of digits exceeds precision. Default precision is 1. If precision is specified as 0 and the value to be converted is 0, the result is an empty string ("").

- f specifies the number of digits after the decimal point. IF a decimal point appears, at least one digit appears before it. The value is rounded to the appropriate number of digits. Default precision is 6; if precision is 0 or if the period appears without a number following it, no decimal point is printed. When the number of digits after the decimal point in the value is less than the precision, letter o should be padded to fill columns. For example, the result of String.format("%2.3f", 1.2) will be "1.200."

Appendix II

- s specifies the maximum number of characters to be printed. By default, all characters are printed. When the width is larger than the precision, the width should be ignored.

Unlike the width argument, the precision argument can cause either truncation of the output value or rounding of a floating-point value.

The type argument is the only required format argument; it appears after any optional format fields. The type character determines whether the given value is interpreted as integer, floating-point, or sting. The supported type arguments are:

- d Integer – the output value has the form [-]dddd, where dddd is one or more decimal digits.

- fr Floating-point – The output value has the form [-]dddd.dddd where dddd is one or more decimal digits. The number of digits before the decimal point depends on the magnitude of the number, and the number of digits after the decimal point depends on the requested precision.

- s String – Characters are printed up to the end of the string or until the precision value is reached.

- % The percent character in the format string can be presented by preceding it with another percent character (%%).

Syntax: String.format(format, value)

Example: The following example formats and displays the user's age (previously calculated):

```
// Display age
Dialogs.alert(String.format("You are %d years old", age));
```

insertAt

Description: Creates a string with a specified element and seperator inserted into the original string at a specified position.

Syntax: String.insertAt(string, element, index, seperator)

Example: The following example inserts an item (in variable item) at the front of a list:

```
// Insert new item at front of list
var list=String.insertAt(list, item,0,",");
```

isEmpty

Description: Returns a true if a string is zero-length (empty), and false if not zero-length.

Syntax: String.isEmpty(string)

Example: The following example displays an error message if a variable is not empty:

```
// Check e-mail address is not empty
if (String.isEmpty(email)) {
 Dialogs.alert("E-mail address cannot be left blank!");
}
```

Appendix II

length

Description: Returns the total number of characters in a string. The result is an integer.

Syntax: String.length(string)

Example: The following example displays an error message if a variable is not the desired length:

```
// Check enough digits entered
if (String.length(acn)!=12){
 Dialogs.alert("Yoou must enter the full 12 digit account number !");
}
```

removeAt

Description: Returns a string with a specified element removed.

Syntax: String.removeAt(string, index, seperator)

Example: The following example removes the last item in a list:

```
// Remove last element from list
var list=String.removeAt(list, String.elements(list)-1,",");
```

replace

Description: Returns a new string where all the occurrences of a string are replaced with a new string.

Syntax: String.replace(string, oldsubstring, newsubstring)

Example: The following example replaces all occurrences of "wap" with "WAP":

```
// Replace all "wap" with "WAP"
var string=String.replace(string, "wap", "WAP");
```

Wireless Markup Language (WML)

replaceAt

Description: Replaces a specified element in a string with a new element.

Syntax: String.replaceAt(string, element, index, seperator)

Example: The following example replaces the first item in a list with an updated value:

```
//Update first element
var list=String.replaceAt(list, new_value, 0,",");
```

squeeze

Description: Replaces consecutive series of white spaces in a string with one white space.

Syntax: String.squeeze(string)

Example: The following example removes any extraneous white spaces in a string:

```
// remove extraneous white spaces
var name=String.squeeze(name);
```

subString

Description: Creates a new string from a specified string, using specified start position and length.

Syntax: String.subString(string, startindex, length)

Example: The following example extracts the first sentence of a paragraph (found by looking for a period):

```
// Get first sentence
var sentence=String.subString(paragraph, 0, String.find(paragraph, ".")-1);
```

Appendix II

toString

Description: Converts a value to a string.

Syntax: String.toString(value)

Example: The following example converts three parts of a phone number into strings so as to be able to concatenate them:

//Build displayable phone number
var phone="("+String.toString(p1)+")"+String.tpString(p2)+"-"+String.toString(p3);

trim

Description: Removes all leading and trailing white spaces from a string.

Syntax: String.trim(string)

Example: The following example trims all leading and trailing space from an email address:

// trim address
var email=String.trim(email);

Wireless Markup Language (WML)

URL library

The URL library contains functions used in URL manipulation.

escapeString

Description: Replaces special characters contained in a string with hexadecimal equivalents.

Syntax: URL.escapeString(string)

Example: The following example escapes a URL query string:

```
// Escape query string
var qs=URL.escapeString(qs);
// Append to url
url=url+"?"+qs;
```

getBase

Description: Returns the absolute URL path of the current WMLS file.

Syntax: URL.getBase()

Example: The following example displays the URL of the script file being executed.

```
// Display file being executed
dialogs.alert("Executing "+URL.getBase());
```

Appendix II

getFragment

Description: Returns the fragment (bookmark) portion of a URL. getFragment() supports both absolute and relative URLs.

Syntax: URL.getFragment(url)

Example: The following example returns a URL's fragment:

```
// Get fragment
var frag=URL.getFragment(url);
```

getHost

Description: Returns the host name portion of a URL. getHost() supports both absolute and relative URLs.

Syntax: URL.getHost(url)

Example: The following example displays the name from which the current script being executed was retrieved:

```
// Display file being executed
dialogs.alert("Executing script retrieved from "+URL.getHost(URL.getBase()));
```

Wireless Markup Language (WML)

getParameters

Description: Returns all the parameters from a URL. If URL contains no parameters, the returned string is empty.

Syntax: URL.getParameters(url)

Example: The following example extracts the query string from a specified URL:

// Get parameters
var param=URL.getParameters(url);

getPath

Description: Result is the path portion of a URL.

Syntax: URL.getPath(url)

Example: The following example extracts the path from a specified URL:

// Get path
var path=URL.getPath(url);

Appendix II

getPort

Description: Returns the port portion of a specified URL. If URL does not specify a port, the resule is an empty string.

Syntax: URL.getPort(url)

Example: The following example extracts the port from a URL if it exists.

```
// get port
var port=URL.getPort(url);
// if empty, use default of "80"
 if (port=="") {
 port ="80";
}
```

getQuery

Description: Return the query string associated with URL. If URL does not contain a query string, the result is an empty string.

Syntax: URL.getQuery(url)

Example: The following example extracts the query string from a specified URL:

```
// get query string
var qs=URL.getQuery(url);
```

Wireless Markup Language (WML)

getReferer

Description: Returns the URL that called the current file. If there is no referrer, the result is an empty string.

Syntax: URL.getReferer()

Example: The following example goes back to the referrer:

```
// Go back to referrer
WMLBrowser.go(URL.getReferer());
```

getScheme

Description: Result is the scheme (protocol) of the current

Syntax: URL.getScheme(url)

Example: The following example extracts the scheme and saves it (assigning a default value if empty):

```
// get scheme
var scheme=URL.getScheme(url)
// If empty, use default of "http"
if (scheme=="") {
 scheme="http";
}
```

Appendix II

isValid

Description: Checks URL for the correct syntax. If URL is properly formatted, the result is true. Otherwise, the result is false.

Syntax: URL.isValid(url)

Example: The following example goes to a URL if it is valid and displays an error message if it is not:

```
// check if valid
if (URL.isValid(url)){
 // go to it
 WMLBrowser .go(url);
} else{
// invalid, display error message
 Dialogs.alert(url+" is not a valid URL!")
}
```

loadString

Description: This function will return the content found at URL. You can only specify one content type. The content type must be text, but the subtype can be anything.

Syntax: URL.loadString(url,contenttype)

Example: The following example retrieves the content of a URL and passes it to a function for processing:

```
// get url and process
var content=URL.loadString(url, "text/vnd.wap.wml");
process (content);
```

resolve

Description: Returns an absolute URL by combining a specified base URL and embedded URL. If the embedded URL is already an absolute URL, it is returned as the result of the function.

Syntax: URL.resolve(baseurl, embeddedurl)

Example: The following example resolves the root a specified URL and then goes to it:

```
// get host root
var resolved = URL.resolve(URL.getHost (url1), "/")
// go for it
WMLBrowser.go(resolved);
```

unescapeString

Description: Converts an escaped URL string (such as any URL you might have escaped using URL.escapeString) back into the characters it represents.

Syntax: URL.unescapeString(string)

Example: The following example converts a variable to its unescaped form:

```
// unescape string
var unescaped=URL.unescapeString(escaped);
```

Appendix II

WMLBrowser Library

The WMLBrowser library contains functions used to manipulate the WML browser. All these functions will return invalid if no WML browser is present or it the WMLScript interpreter is not involved by the WML browser.

getCurrentCard

Description: Result is the shortest relative URL of the current WML card.

Syntax: WMLBrowser.getCurrentCard()

Example: The following code returns the card that is currently at the top of the history stack:

```
// get current card
function browserCard(){
 var curCard = WMLBrowser.getCurrentCard();
}
```

getVar

Description: Access the value of the variable specified by name in the browser.

Syntax: WMLBrowser.getVar(name)

Example: The following code gets the value of the variable "zip" out of the WML variable stack:

```
// get "zip" variable
var zipvalue=WMLBrowser.getVar("zip");
```

Wireless Markup Language (WML)

go

Description: Performs the same task as the <go> element in WML. The WMLBrowser.go function is executed only after WMLS rendering is complete. You can have multiple calls to WMLBrowser.go, but each one is overriding. If the last call to this function had an empty string for the value of URL, no content would be loaded.

Syntax: WMLBrowser.go(url)

Example: The following example goes to a URL if it is valid, and displays an error message if it is not:

```
// check if valid
if (URL.isValid(url)){
// go to it
 WMLBrowser.go(url);
}
else {
// invalid display error message
 Dialogs.alert(url +" is not a valid URL");
}
```

newContext

Description: This function operates in the same manner as the WML attribute newcontext. It erases the browser context and returns an empty string as its result.

Syntax: WMLBrowser.newContext()

Example: The following code causes the browser context to be erased:

```
// create new context
WMLBrowser.newContext();
```

Appendix II

prev

Description: Forces a WML <prev> task, it takes the browser back to the previous WML card. This function is executed only after all WMLS interpretation is finished.

Syntax: WMLBrowser.prev()

Example: The following code causes the browser to navigate one step backward in the history stack:

```
// go back one step
WMLBrowser.prev();
```

refresh

Description: Works the same as the WML refresh task. This function will force the browser to update its context and use the latest version of retrieved cards.

Syntax: WMLBrowser.refresh()

Example: The following code causes the current card to be refreshed:

```
// check if refresh needed
if (need_refresh) {
 WMLBrowser.refresh();
}
```

Wireless Markup Language (WML)

setVar

Description: The variable with the specified name is set to contain the given value in the current browser context. Variable name and its vakue must follow the syntax specified by WML.

Syntax: WMLBrowser.serVar(name, value)

Example: The following code sets the value of "done" in a variable named "status":

```
// set "status"
WMLBrowser.setVar("status", "done");
```

Dialogs Library

The Dialogs library contains functions used to create user dialogs.

alert

Description: Allows you to send message text to the user's device. Once the user confirms receipt of the message, an empty string is returned.

Syntax: Dialogs.alert(message)

Example: The following code expects a string input. It checks the length of the string, and if it is fewer than 8 characters, the message "Login must be longer than 8 chars" is displayed on the device screen:

```
// check login length
if (String.length(login) > 8 ) {
// too short, ask again
 Dialogs.alert("Login must be longer than 8 chars");
}
```

Appendix II

confirm

Description: Allows you to send a message to the user's device. The user is given two selection options, ok or cancel (the text for each is specified). If the user chooses ok, the result of the function is true. Otherwise, the result is false.

Syntax: Dialogs.confirm(message, ok, cancel)

Example: The following example displays the string "Confirm your order" on the device display, and assigns a value of true to the variable named "confirm" if the user presses the accept key, and a value of false if he presses the soft1 key.

```
// get order confirmation
var confirm= Dialogs.confirm("Confirm your order", "Yes", "No");
```

prompt

Description: Allows you to send message to the user's device prompting for input. Default input text may be provided.

Syntax: Dialogs.prompt(message, defaultinput)

Example: The following code displays the string "Enter phone number" on the device display, along with an input element that contains a default value of "650-". When the user presses the accept key, the value is assigned to the variable named "phone":

```
// get phone number
var phone = Dialogs.prompt("Enter phone number", "650-");
```

Debug Library

The Debug library is a Nokia-specific library containing debugging functions.

closeFile

Description: Closes an open debug file.

Syntax: Debug.closeFile()

Example: The following example opens an existing file, writes log text to it, and then closes it:

```
// create output file
Debug.openFile("C:\\log\\log.txt","a");
// write debug output
Debug.printLn("Value of x is: " + x);
// close log file
Debug.closeFile();
```

Appendix II

openFile

Description: Opens a file for reading or writing, used primarily for embedding trace information in scripts. Function supports three modes, "r" for reading, "w" for writing (overwriting existing file if present), and "a" for appending to an existing file.

Syntax: Debug.openFile(filename, mode)

Example: The following example opens an existing file, writes log text to it, and then closes it:

```
// create output file
Debug.openFile("C:\\log\\log.txt","a");
// write debug output
Debug.printLn("Value of x is: " + x);
// close log file
Debug.closeFile();
```

printLn

Description: Writes specified debug output to an already opened file.

Syntax: Debug.printLn(string)

Example: The following example opens an existing file, writes log ttext ot it, and then closes it:

```
// create output file
Debug.openFile("C:\\log\\log.txt","a");
// write debug output
Debug.printLn("Value of x is: " + x);
// close log file
Debug.closeFile();
```

Console Library

The Console library is a Phone.com specific library containing debugging functions.

print

Description: Writes debug text to the phone information window; no new line character is inserted after the text.

Syntax: Console.print()

Example: The following example writes debug text to the phone information window:

```
// write debug output
Console.print("In loop, iteration: " + String.toString(i));
```

printLn

Description: Writes debug text to the phone information window and inserts a new line character after the text.

Syntax: Console.printLn()

Example: The following example writes debug text to the phone information window:

```
// write debug output
Console.printLn("In loop, iteration: " + String.toString(i));
```

Appendix III

Toolkits Available On-line

This appendix contains the website address and a brief description of some of the toolkits currently available online.

Java Runtime Environment

Website: http://java.sun.com/j2se/downloads.html

Description of Tool:

> The Java 2 SDK includes tools useful for developing and testing programs written in the Java programming language and running on the Java platform. These tools are designed to be used from the command line. Except for the appletviewer, these tools do not provide a graphical user interface.
>
> The on-line Java 2 Platform Documentation contains API specifications, feature descriptions, developer guides, reference pages for SDK tools and utilities, demos, and links to related information. This documentation is also available in a download bundle which you can install on your machine.

Wireless Markup Language (WML)

Nokia WAR Toolkit

Website: http://www.forum.nokia.com

Description of tool:

This package contains Nokia Mobile Internet Toolkit 4.0 (NMIT 4.0), Nokia Update Manager 1.1 (NUM 1.1) and Nokia Content Publishing Toolkit 2.0.1 (NCPT 2.0.1). NMIT is a set of editors for creating various types of mobile Internet content such as WML, WMLScript, xHTML, MMS and WAP that can be previewed on supported phone SDKs. A range of Content Authoring SDKs are available via separate downloads, that can be used with NMIT for previewing content. NUM 1.1 shows the various Nokia products installed on your machine and provides details about available updates and relationship between products. NCPT 2.0.1 allows you to generate OMA DRM content packages and related download descriptors.

System Requirements:

* Windows 2000 (SP 2) or Windows XP
* JRE 1.4.1 is required (1.4.1_02 is recommended)

Appendix III

Nokia Developer Platform

Website: http://www.forum.nokia.com/main/1,6566,010,00.html

Description of Tool:

Nokia's Developer Platform approach is designed to help developers build and deliver mobile applications to a global audience in less time, with less effort and cost. Developers can build core functionality on top of the Platform technologies, and then optimize their applications for specific user interfaces and technology extensions for target devices.

Developer Platform 1.0 for Series 40 brings color displays, Java™ programmability, and multimedia messaging to the mass market. It provides a rich, common set of Java APIs, a browsing environment based on the WAP 1.2.1 specification, and multimedia messaging. Nokia projects that 50 million to 100 million consumers around the world will own Series 40 devices by the end of 2003. For more information about Developer Platform 1.0 for Series 40 and the tools available, please visit www.forum.nokia.com/series40.

The Series 60 Platform brings open standards and multi-vendor support to the smartphone market. Built on Symbian OS, the Series 60 Platform provides a rich, common set of Java APIs, full C++ programmability, browsing, and the multimedia messaging environment that developers need to build everything from rich games to secure enterprise applications. Nokia projects that more than 10 million Series 60 devices will be in use globally by the end of 2003. For more information about the Series 60 Platform and the tools available, please visit www.forum.nokia.com/series60.

The Developer Platform for Series 40 and Series 60 offers sustainable business opportunities for mobile application developers. The Developer Platform gives developers the ability to reach consumers and enterprises alike on a mass-market scale. In addition to providing volume benefits, Platform lets developers easily keep pace with new software and hardware technologies.

Wireless Markup Language (WML)

Openwave Phone Simulator

Website: http://developer.openwave.com/omdt/client_sdk.html

Description of Tool:

The Openwave Phone Simulator is a free software development kit that makes creating innovative applications for the mobile Internet easier than ever before. Featuring mobile phone simulators, this flexible and powerful programming tool also includes the latest versions of the Openwave® Mobile Browser and Openwave® Mobile Messaging client, allowing application developers to test and demonstrate their software while still in the development cycle. Developers can use the simulator to validate their mobile applications that generate XHTML Mobile Profile / CSS, MMS-SMIL, WAP Push, WML or cHTML. Openwave Phone Simulator includes debugging tools, sample code and documentation to help developers create powerful applications for the Openwave platform.

Pixo Mobile Download Server

Website: http://www.pixo.com/products/products003.html

Description of Tool:

Developing a profitable download service requires more than just downloading a Java application to a device. Profitability comes from creating a wireless Java ecosystem where operators, content providers, enterprises and wireless ASPs all share revenue and responsibility for success. At the center of this ecosystem, the operator's primary focus is on creating a shared revenue model, forming relationships with content providers to ensure a wide selection of content, and providing a simple user experience. With a

Appendix III

wireless Java solution that leverages existing operator infrastructure and serves many subscribers on a minimum of hardware, operators can achieve compelling ROI with a small upfront investment.

The Pixo Mobile Download Server (MDS) is a high performance software solution that allows operators to manage the complexity of delivering content over-the-air (OTA) to their subscribers. Pixo MDS separates content aggregation, presentation, and delivery into separate Managers. This enables an operator to centrally manage content, customize multiple subscriber interfaces, and ensure fast downloads.

To support the widest range of downloadable content, Pixo MDS now includes the Extensible Content Framework (ECF), which enables the Operator with simple configuration, rather than Pixo or a systems integrator, to directly add support for new downloadable content, legacy content, and other custom types without interrupting the download service. Today, subscribers are using MDS to download wireless Java applications, ring tones (polyphonic), pictures, Symbian applications, Microsoft Smartphone, Pocket PC applications, video clips and other custom Operator-specific content.

The Pixo Mobile Download Server is built upon the Java 2 Enterprise Edition (J2EE)™ Framework for carrier-grade scalability and robustness. An easy to use, Web-based interface allows developers to submit applications for review by the operator. The server's application management and download systems provide the tools to approve, categorize and deliver the applications to subscribers. With distributed downloading, operators can configure multiple Vending Managers (e.g. "Vending Machines") to integrate with a central, backend warehouse of listed and published content through the Catalog Manager. Download and Discovery adapters let operators implement different discovery and download protocols as needed to support new devices and a variety of content types, including ring tones, images, Java and non-Java applications, and Operator-specific content types. The server also provides the set of critical APIs needed for deployment in the operator environment.

Wireless Markup Language (WML)

Index

Acceptcharset, 62
Access, 2, 4, 9, 11, 14-15, 19, 21, 25-26, 28-29, 40, 103-105, 108, 117, 138, 141, 176-177, 179, 182, 200, 202-204, 207-208, 228
Acterbased, 179
Adding Security to Applications, 191
Advanced WML Script, 143
Alerts, 68, 182-184, 188
Anchor, 68, 70, 72, 78-79, 214, 235, 246
Apache, 175
API, 176
Applet, 34
Array, 148
ASCII, 50
Attributes, 45-46, 56-58, 64, 66, 74-75, 78, 83, 86, 88, 93, 98, 103-104, 106, 109, 203, 215, 217, 229, 236, 248, 250
Bandwidth, 4, 11, 233
Basic Commands, 56
Basic Structure, 43, 79
Bearer, 176-177, 179
Bluetooth, 29, 33, 35
BREW, 34, 41-42
Bring, 182
Broadcasts, 4
Browsing, 29-30, 193, 226
Build, 199
Builder, 37

Cache, 104, 182-185, 188
Cache Control, 183
Cache Operations, 182-183, 188
Calling WMLScript Functions, 137-138
Card, 44-45, 56-62, 66, 75, 77-79, 81-84, 88, 98, 100, 103, 106, 108-110, 112-113, 142, 194
CDMA, 4, 38
Cellular and PCS, 1
CGI, 46, 62, 64, 105, 107, 161, 226, 228-229, 246
CGIBIN, 209, 233, 245
Changing Text Color, 238
Checkbox, 229, 244, 246
cHTML, 8-10, 19-20, 22-23, 26, 209, 233-236, 238-239, 241, 244-251
Clickable, 228
Client, 181, 184, 186-187, 195, 200-202, 207
Client Authentication, 201-202
Console, 146, 170-172
Content Messages, 183
Content Push, 181
Context, 15, 201
Convergence Devices, 25, 30, 32, 40, 42
Cookies, 19, 233, 246
Creating Push and Pull Notifications, 181

267

Wireless Markup Language (WML)

Cryptography, 193, 199-200, 205-206
Datagram, 176
Debug, 146, 170-172
Decks, 43, 50, 77, 88, 90, 104, 107, 183, 203-204, 207
Declaration, 111, 113, 115, 117, 126, 139
Dialogs, 112, 139, 146, 167-169, 171-173
Digits, 66, 116, 120, 139
Do, 6, 15, 32, 42, 59-61, 77-79, 83-84, 109, 141, 143, 146, 148, 170, 172-173, 180, 182-184, 199, 201, 204, 208, 217, 248
Domain, 104-105, 117, 138, 203-204
Download, 5, 21, 26, 34, 183, 186
Drivers, 250
Dropdown, 5, 229
Dualmode, 26, 42
Ebusiness, 4
ECMAscript, 111
Effect, 75, 129, 220
Electroluminescent, 26
Embed, 218
Emulator, 178
Encrypting, 192
Encryption, 192-193
Enumerated, 224
EPOC, 35, 41-42
Escape String, 161, 163, 173
Escapement, 50
Event, 60-61, 98, 106, 108, 112, 181, 184, 187-188, 204
Ex51, 114, 137
Ex55init, 82
Explorer, 38
FDM, 4, 21
FDMA, 4
Fetch, 186
Field, 57, 64-66, 68, 78, 101, 103, 105, 108-110, 120
Fieldset, 101, 103, 108-110
Firewall, 200
Float, 112, 117, 120, 139, 146, 158, 160, 171, 173
Floatingpoint, 125, 143-144, 154, 156, 158, 160, 163, 171
Font, 19, 22, 51, 77, 233, 238-239, 246-247
FTP, 178-179
Func, 142, 173
Functions, 26, 30, 35, 42, 46, 88, 90, 98, 111-114, 119-120, 126, 137-138, 140, 143, 146, 148, 152, 154, 156, 158, 161, 164-165, 167-173, 180
Gateways, 182, 188, 200
Getvar, 120, 164-166
GIF, 88, 107, 177-178, 215, 236, 245, 247
GPRS, 38
GSM, 2-3, 21, 23, 30, 35, 38, 176
HDML, 9
Head, 103, 105, 108, 203, 210-211, 233, 245
Headers, 212, 233, 245, 248
Hidden, 228, 244, 246, 250
Hierarchical, 62, 96, 108, 224, 242
Hijacked, 200
Href, 61, 70, 114, 117, 120, 214, 235, 245
Hspace, 86
HTM, 210, 233
HTML, 9-10, 14-16, 18-19, 22, 30, 43, 46, 51-52, 54, 70, 72, 86, 88, 108, 111, 138, 177, 179, 193, 196, 199,

Index

202, 205, 207, 210-211, 215, 217-218, 233, 238, 244-246
HTTP, 11-14, 16, 22, 62, 104, 117, 175-177, 179, 193, 198-199, 202-204, 207, 210, 228
Hyperlink, 43, 57, 68, 78, 214-215, 245
Hypertext, 9, 17, 19, 22
IETF, 202
IFELSE, 124, 126, 140-141
Imagelink, 217
Images, 9, 14, 19, 43, 70, 72, 86, 88, 107, 178-179, 214-215, 217, 236, 245, 247, 249
IMG, 86, 107, 215-217, 236, 245, 247-248
Imode, 9-10, 19, 26, 209, 233, 236
Input, 11-12, 49-50, 64-68, 78-79, 81, 88, 90, 101, 105, 107, 114, 120, 148, 174, 228-229, 231, 233, 244, 246, 248-250
Inserting Comments, 46
Integer, 93, 100, 111, 115-116, 119, 125, 143-144, 148, 150, 154, 156, 158, 160, 163, 171, 173
Interlaced, 236
Interoperable, 11
Intranet, 28
IP, 12, 35, 176, 193
Irda, 33
IS136, 4
Isempty, 152
Isfloat, 156
Isn, 141
Iso, 201
Isp, 19, 191
Isvalid, 117, 125, 140-141, 144
Java, 26, 64, 116, 131, 161, 176, 209, 233, 245
Javascript, 15-16, 22, 111-112, 141
JPEG, 19, 88, 107, 177-178, 215, 245
Lang, 112, 119, 139, 146, 154, 156, 171, 173
Libraries, 112-113, 117, 138-141, 143, 146, 148, 171, 173
Link, 2, 7, 57, 68, 70, 82, 100, 124, 154, 171, 181-182, 186, 203-204, 214-215, 235-236, 246, 250
Localsrc, 86, 109
Login, 202
Loop, 82, 126, 129, 131, 133, 135
Maxfloat, 120
Mbusiness, 4
Menu, 183
Meta, 103, 108-109, 117
Meta, 103, 108-109, 117
Microbrowser, 12, 25, 38-39, 41-42, 177, 210, 220
MicroBrowsers, 12, 25, 38, 41, 177, 220
Minfloat, 120
MMS, 26, 36
Mobile Data Terminals, 25, 27, 40, 42
Mobile Devices, 18, 25-26, 32, 34, 37, 40
Mobile Operating Systems and Platforms, 34
Mobile Phones, 6-7, 25-26, 30, 32, 35, 37, 40-42, 88, 178
Mobitex, 27
Modifying an HTTP Server, 177

More Complex XHTML Basic Forms, 229
MP3, 30
MSC, 2
Multimedia, 25-26, 35-36
Multi-part Messages, 182-183, 188
Multiplexing, 2, 4, 21
Narrowband, 15
Nested and Ordered Lists, 224, 242
Nesting, 17, 22, 32, 46, 57, 70, 72, 82, 112-114, 117, 124, 126-127, 129, 135, 141, 187, 201, 204, 210-211, 220, 224, 226, 228, 242, 244, 246-248, 250
Newcontext, 57-58
Next Generation, 32, 35, 41
Nokia WAP Server, 175-176, 179-180
Nonascii, 177
Nonmatch, 142
Nonnegative, 93
Nonrepudiation, 192-194, 199, 201, 205
Nonvisual, 228
Noop, 77-78, 83-84
Noop, 77-78, 83-84
Nowrap, 54
OFDM, 2, 4, 21
Onevent, 83, 106-109
Operators, 11, 112, 122-125, 139-141, 144, 171, 173
Optgroup, 92, 96, 98, 107-110
Option, 68, 92-93, 95-96, 107, 109, 229, 244
OSS, 34, 38
Other Script Languages, 209
Packet Switching , 5
Paging, 6-8, 21
Paging, 6-8, 21

Passing Variables, 88
Personal Digital Assistant (PDA) , 1, 5, 8, 11, 25, 29-32, 35, 37, 40, 42, 52, 101
Picocells, 2
Pixels, 86, 88, 215, 236
Platform, 25, 32, 34-35, 37, 40-41, 57, 196, 199-200, 207
Pocket PC, 30-33, 37
Postfield, 61, 105, 108
Pragma, 138
Presence, 32, 218
Prev, 46, 68, 75-76, 78, 81, 84, 88
Profile, 10
Programmability, 15, 22
Programming, 10, 14, 19, 35, 43, 46, 81, 105, 111, 233, 248
Protocol, 8, 11, 14, 16, 22, 27, 35, 43, 61, 176, 178-180, 198-199, 206, 228
Proxy, 184, 186, 197-199
PSION, 35, 41
Public Key, 193, 199-201, 205-206
Public Switched Telephone Network (PSTN), 2, 7
Publishing to the WAP Server, 178
Pull Notification, 184, 186, 188-189
Pull, 181-182, 184, 186-189
Push Notification, 184-185
Push, 48-49, 60-62, 65, 81, 84, 88, 93, 112, 167, 181-182, 184-189, 196, 226
Refresh, 46, 68, 88, 98-99, 106, 108-110, 146, 167, 204
Request Path, 196
Retransmitting, 191
Roaming, 3, 27
RSA, 199
Runtime, 34, 41
Scripting Structure, 111

Index

Scripting, 15-16, 19, 22, 111, 139, 228, 233, 246
Second Generation
　　　　Security, 62, 66, 142, 191-196, 198-201, 204-208
Security Basics, 191
Security Certificates, 200, 207
Select, 26, 62, 64, 92-93, 95, 101, 107, 229, 231, 244, 246
Separator, 146
Server based, 105
Server, 11-14, 16, 18-19, 23, 43, 46, 61-62, 64, 81, 86, 104-105, 108, 137, 140, 175-188, 191, 193, 196-199, 202-203, 207, 226, 228, 246
Servlet, 64, 105, 161, 176-177, 179
Session Management, 201
Session, 191, 196-197, 201, 206
Setting up a WAP Server, 175
Setvar, 46, 48, 61, 88, 90, 98, 106-107, 109, 114, 117, 146, 167, 204
SGML, 218
Simple XHTML Basic Forms, 226
Smartphone, 36-37
Special Characters, 50, 120, 161, 218-219, 239, 245, 247
Special Characters and More Line Breaks, 239
SSL, 193, 199, 205-208
Standard Libraries, 112, 139, 141, 143, 146, 171, 173
Statements and Expressions, 126
Stinger, 37-38, 41-42
Stinger OS, 37, 41
Streaming, 5, 21, 32, 35
String, 64, 111-113, 115-117, 122, 125-126, 139, 142-144, 146, 148, 150, 152, 161, 163, 171, 173-174
Symbian, 30, 35-36, 41
Table, 9-10, 16, 45, 59, 66, 72-75, 78-79, 82, 178, 220, 222, 246, 248, 250
Tables, 17, 19, 22, 72, 210, 220, 233, 246, 248
TDMA, 2-4, 21
Templates, 83
Text Formatting, 51, 77, 233
Threat Models, 194
Timer, 57, 84, 100, 108-110
Title, 57-58, 68, 70, 74, 79, 93, 95-96, 101, 142, 210-211, 233
Tones, 26
Toolkit, 44, 57, 88, 95, 101, 110, 178, 210, 233
Trusted, 191-194, 201
Two-Way Pagers, 25, 29, 40, 42
Type Conversion of Variables, 50, 143, 171, 173
Type Conversions, 143, 145, 154, 171
Unordered Lists, 222, 224, 241-242, 246-249
URL, 13, 43, 50, 57, 61-62, 70, 77-79, 84, 86, 96, 104, 111-113, 117, 138-139, 146, 161, 163, 166, 171, 173, 177, 179, 182, 185-186, 188, 228
URL, 13, 43, 50, 57, 61-62, 70, 77-79, 84, 86, 96, 104, 111-113, 117, 138-139, 146, 161, 163, 166, 171, 173, 177, 179, 182, 185-186, 188, 199, 228
Validation, 161, 173
Variables, 46-47, 50, 57, 68, 70, 75, 77-79, 88, 90, 98, 105-107, 109-110, 115-117, 120, 124, 126, 139, 141, 143-144, 163-165, 167, 171, 173-174, 198, 204, 207-208
WAP

271

Wireless Markup Language (WML)

Gateway, 12-13, 137, 175-176, 179, 182, 184-187, 191, 196-201, 203, 206-208
Wapbased, 183, 188
Wapcompliant, 146
Wbmp, 88, 107, 110, 178-179, 183, 188, 215
Wbxml, 198
What are Notifications, 181
Windows CE, 30-32, 37, 41-42
Wireless Data, 4-6, 27
Wireless Devices and Operating Systems, 25
Wireless Languages, 8
Wireless Systems and Applications, 1
WLAN, 33
WML Browser, 164, 191, 199
WML Browser, 112, 114, 117, 120, 139, 146, 164, 166-167, 171, 191, 199
WML for Secure Applications, 203
WML Objects and Syntax, 83
WMLScript, 8, 10, 12, 15-17, 22-23, 46, 64, 90, 98, 107, 111-117, 119-120, 122-127, 129-130, 132-133, 137-146, 148, 154, 163-167, 171-174, 178, 183, 188, 209-210, 245
WTLS, 199-201, 206, 208
WTLS and SSL, 199
XHTML, 8-10, 12, 17-19, 22-23, 38, 41, 175, 209-212, 214-216, 218-220, 222, 224, 226, 229, 233, 241, 244-250
XML, 9, 17, 22, 43-44, 198, 210, 245

Wireless Books
by ALTHOS Publishing

Wireless Systems

ISBN: 0-9728053-4-6 **Price:** $34.99
Authors: Lawrence Harte, Dave Bowler, Avi Ofrane, Ben Levitan
#Pages 368 **Copyright Year:** 2004

Wireless Systems; Cellular, PCS, 3G Wireless, LMR, Paging, Mobile Data, WLAN, and Satellite explains how wireless telecommunications systems and services work. There are many different types of wireless systems competing to offer similar types of voice, data, and multimedia services. This book describes what the functional parts of these systems are and the basics of how these systems operate. With this knowledge,

Wireless Dictionary

ISBN: 0-9746943-1-2 **Price:** $39.99
Author: Althos **#Pages:** 628 **Copyright Year:** 2004

The Wireless Dictionary is the Leading wireless industry resource. The Wireless Dictionary provides definitions and illustrations covering the latest voice, data, and multimedia services and provides the understanding needed to communicate with others in the wireless industry. This book is the perfect solution for those involved or interested in the operation of wireless devices, networks, and service providers.

Introduction to 802.11 Wireless LAN (WLAN)

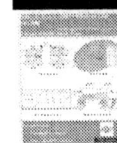

ISBN: 0-9746943-4-7 **Price:** $14.99
Author: Lawrence Harte **#Pages:** 52 **Copyright Year:** 2004

Introduction to 802.11 Wireless LAN (WLAN), Technology, Installation, Setup, and Security book explains the functional parts of a Wireless LAN system and their basic operation. You will learn how WLANs can use access points to connect to each other or how they can directly connect between two computers. Explained is the basic operation of WLAN systems and how the performance may vary based on a variety of controllable and uncontrollable events. This book will explain the key differences between the WLAN systems.

Introduction To Wireless Systems

ISBN: 0-9742787-9-3 **Price:** $11.99
Author: Lawrence Harte, **#Pages:** 68 **Copyright Year:** 2003

Introduction to Wireless Systems book explains the different types of wireless technologies and systems, the basics of how they operate, the different types of wireless voice, data and broadcast services, key commercial systems, and typical revenues/costs of these services. Wireless technologies, systems, and services have dramatically changed over the past 5 years. New technology capabilities and limited restrictions are allowing existing systems to offer new services.

Althos Publishing, 404 Wake Chapel Road, Fuquay NC 27526 USA
1-919-557-2260 1-800-227-9681 Fax 1-919-557-2261 WWW.AlthosBooks.com

Worlds Largest Onlline Wireless Dictionary
WWW.WirelessDictionary.com

Introduction to Paging Systems

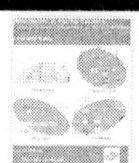

ISBN: 0-9746943-7-1   Price: $14.99
Author: Lawrence Harte   #Pages: 48   Copyright Year: 2004

Introduction to Paging Systems describes the different types of paging systems, what services they can provide, and how they are changing to meet new types of uses. This book explains the different types of paging systems and how they are changing. Explained is how and why paging systems are transitioning from one-way systems to two-way systems.

Introduction to Satellite Systems

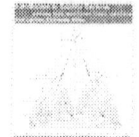

ISBN: 0-9742787-8-5   Price: $11.99
Author: Ben Ievitan, Lawrence Harte   #Pages: 48   Copyright Year: 2004

In 2003, the satellite industry was a high-growth business that achieved over $83 billion in annual revenue. This book offers an introduction to existing and soon to be released satellite communication technologies and services. It covers how satellite systems are changing, growth in key satellite markets and key technologies that are used in satellite systems.

Introduction to Mobile Data

ISBN: 0-9746943-9-8   Price: $14.99
Author: Lawrence Harte   #Pages: 628   Copyright Year: 2004

Introduction to Mobile Data explains how people use devices that can send data via wireless connections, what systems are available for providing mobile data service, and the services these systems can offer. This book explains the basics of circuit switched and packet data via wireless mobile systems. Included are descriptions of various public and private systems that are used for data and messaging services.

Introduction to Private Land Mobile Radio

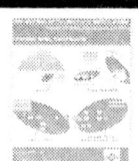

ISBN: 0-9746943-6-3   Price: $14.99
Author: Lawrence Harte   #Pages: 50   Copyright Year: 2004

Introduction to Private Land Mobile Radio explains the different types of private land mobile radio systems, their basic operation, and the services they can provide. This book covers the basics of private land mobile radio systems including traditional dispatch, analog trunked radio, logic trunked radio (LTR), and advanced digital land mobile radio systems. Described are the basics of LMR technologies including simplex, and half-duplex.

Introduction to GSM Systems

ISBN: 1-9328130-4-7   Price: $14.99
Author: Lawrence Harte   #Pages: 48   Copyright Year: 2004

Introduction to GSM describes the fundamental components, key radio and logical channel structures, and the basic operation of the GSM system. This book explains the basic technical components and operation of GSM technology. You will learn the physical radio channel structures of the GSM system along with the basic frame and slot structures.

Althos Publishing, 404 Wake Chapel Road, Fuquay NC 27526 USA
1-919-557-2260 1-800-227-9681 Fax 1-919-557-2261 WWW.AlthosBooks.com

Worlds Largest Online Wireless Dictionary
WWW.WirelessDictionary.com

Internet Telephone Basics

ISBN: 0-9728053-03 Price: $29.99
Author: Lawrence Harte #Pages:226 Copyright Year: 2003

Internet Telephone Basics explains how and why people and companies are changing to Internet Telephone Service. Learn how much money can be saved using Internet telephone service and how you can to use standard telephones and dial the same way. It describes how to activate Internet telephone service instantly and how to display your call details on the web.

Voice Over Data Networks for Managers

ISBN: 0-9728053-2-X Price: $49.99
Author: Lawrence Harte #Pages:352 Copyright Year: 2003

Voice over Data Networks for Managers explains how to reduce communication costs 40% to 70%, keep existing telephone systems, and ways to increase revenue from new communication applications. Discover the critical steps companies should take and risks to avoid when transitioning from private telephone systems (KTS, PBX, and CTI) to provide new services.

Patent or Perish

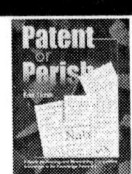

ISBN: 0-9728053-3-8 Price: $39.95
Author: Eric Stasik #Pages:220 Copyright Year: 2003

Patent or Perish Explains in clear and simple terms the vital role patents play in enabling high technology firms to gain and maintain a competitive edge in the knowledge economy. Patent or Perish is a Guide for Gaining and Maintaining Competitive Advantage in the Knowledge Economy. In a world where knowledge has value and knowledge creates value, ideas are the new source of wealth. This book describes how technologies like the Internet remove traditional barriers to entry and enable competitors to quickly.

Telecom Basics 3rd Edition

ISBN: 0-9728053-5-4 Price: $34.99
Author: Lawrence Harte #Pages:356 Copyright Year: 2003

This introductory book provides the fundamentals of signal processing, signaling control, can call processing technologies that are used in telecommunication systems. Covered are the key facets of voice and data communications, ranging from such basics as to how a telephone set works to more complex topics as how to send voice over data networks and the ways calls are processed in public and private telephone systems.

Introduction to Cable Television Systems

ISBN: 0-9728053-6-2 Price: $12.99
Author: Lawrence Harte #Pages: 48 Copyright Year: 2004

Community access television (CATV) is a television distribution system that uses a network of cables to deliver multiple video, data, and audio channels. This excerpted chapter from Telecom Systems provides an overview of cable television systems including cable modems, digital television, high definition television (HDTV), and the market growth of cable television and advanced services such as video on demand.

Althos Publishing, 404 Wake Chapel Road, Fuquay NC 27526 USA
1-919-557-2260 1-800-227-9681 Fax 1-919-557-2261 WWW.AlthosBooks.com

Worlds Largest Online Wireless Dictionary
WWW.WirelessDictionary.com

Introduction to Private Telephone Networks 2nd Edition

ISBN: 0-9742787-2-6   Price: $12.99
Author: Lawrence Harte   #Pages: 48   Copyright Year: 2004

Private telephone networks are communication systems that are owned, leased or operated by the companies that use these systems. They primarily allow the interconnection of multiple telephones within the private network with each other and provide for the sharing of telephone lines from a public telephone network.

Introduction to Telecom Billing

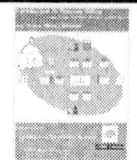

ISBN: 0-9742787-4-2   Price: $11.99
Author: Lawrence Harte   #Pages: 36   Copyright Year: 2003

This book explains how companies bill for telephone and data services, information services, and non-communication products and services. Billing and customer care systems convert the bits and bytes of digital information within a network into the money that will be received by the service provider. To accomplish this, these systems provide account activation and tracking, service feature selection, selection of billing rates for specific calls, invoice creation, payment entry and management of communication with the customer.

Introduction to Public Switched Telephone Networks 2nd Edition

ISBN: 0-9742787-6-9   Price: $34.99
Author: Lawrence Harte   #Pages: 48   Copyright Year: 2004

Public telephone networks are unrestricted dialing telephone networks that are available for public use to interconnect communications devices. There are also descriptions of many related topics, including: Local loops, switching systems, numbering plans, market growth, public telephone system interconnections, and common channel signaling (SS7),

Introduction to SS7 & IP Telephony

ISBN: 0-9746943-0-4   Price: $14.99
Author: Lawrence Harte   #Pages: Copyright Year: 2004

The Introduction to Signaling System 7 (SS7) and IP control system that is used in public switched telephone networks (PSTN) can be interconnected to other types of systems and networks using Internet Protocol (IP). Some of the interconnection issues relate to how the control of devices can be performed using dissimilar systems.

Introduction to IP Telephony

ISBN: 0-974278-7-7   Price: $12.99
Author: Lawrence Harte   #Pages: Copyright Year: 2003

This "Introduction to IP Telephony" book explains why companies are converting some or all of their telephone systems from dedicated telephone systems (such as PBX) to more standard IP telephony systems. These conversions allow for telephone bill cost reduction, increased ability to control telephone services, and the addition of new telephone information

Althos Publishing, 404 Wake Chapel Road, Fuquay NC 27526 USA
1-919-557-2260 1-800-227-9681 Fax 1-919-557-2261 WWW.AlthosBooks.com

Worlds Largest Onlline Wireless Dictionary
WWW.WirelessDictionary.com

Signaling System Seven (SS7) Basics 3rd Edition

ISBN: 0-9728053-7-0   Price: $34.99
Author: Lawrence Harte   #Pages: 276   Copyright Year: 2003

This introductory book explains the operation of the signaling system 7 (SS7) and how it controls and interacts with public telephone networks and VoIP systems. SS7 is the standard communication system that is used to control public telephone networks. In addition to voice control, SS7 technology now offers advanced intelligent network features and it has recently been updated to include broadband control capabilities.

Introduction to SIP IP Telephony Systems

ISBN: 0-9728053-8-9   Price: $14.99
Author: Lawrence Harte   #Pages: 117   Copyright Year: 2004

This book explains why people and companies are using SIP equipment and software to efficiently upgrade existing telephone systems, develop their own advanced communications services, and to more easily integrate telephone network with company information systems. This book provides descriptions of the function parts of SIP systems along with the fundamentals of how SIP systems operate.

Telecom Systems

ISBN: 0-9728053-9-7   Price: $34.99
Author: Lawrence Harte   #Pages: 480   Copyright Year: 2004

This book Telecom Systems shows the latest telecommunications technologies are converting traditional telephone and computer networks into cost competitive integrated digital systems with yet undiscovered applications. These systems are continuing to emerge and become more complex. Telecom Systems explains how various telecommunications systems and services work and how they are evolving to meet the needs of bandwidth

Introduction to Transmission Systems

ISBN: 0-9742787-0-X   Price: $14.99
Author: Lawrence Harte   #Pages: 52   Copyright Year: 2004

This book explains the fundamentals of transmission lines and how radio waves, electrical circuits, and optical signals transfer information through a communication medium or channel on carrier signals. It also explains the ways that a single line can be divided into multiple channels and how signals are carried over transmission lines in analog or digital form.

Tehrani's IP Telephony Dictionary

ISBN: 0-9742787-1-8   Price: $39.99
Author: Althos   #Pages: 628   Copyright Year: 2003

Tehrani's IP Telephony Dictionary, The Leading VoIP and Internet Telephony Resource provides over 10,000 of the latest IP Telephony terms and more than 400 illustrations to define and explain latest voice over data network (VoIP) technologies and services. It provides the references needed to communicate with others in the communication industry.

Althos Publishing, 404 Wake Chapel Road, Fuquay NC 27526 USA
1-919-557-2260 1-800-227-9681 Fax 1-919-557-2261 WWW.AlthosBooks.com

Worlds Largest Onlline Wireless Dictionary
WWW.WirelessDictionary.com

Practical Patent Strategies Used by Successful Companies

ISBN: 0-9746943-3-9 Price: $14.99
Author: Eric Stasik #Pages: Copyright Year: 2004

This book explains how companies can use patent strategies to achieve their business goals. Patent strategies may be considered abstract legal or economic concepts. Examining how patents are used by leading companies in specific business applications can provide great insight to their practical use and application in your business plan. This book presents in plain and clear language why having a patent strategy is important.

Introduction to xHTML

ISBN: 0-9328130-0-4 Price: $34.99
Author: Lawrence Harte #Pages: Copyright Year: 2004

This book explains what is xHTML Basic, when to use it, and why it is important to learn. You will discover how the xHTML Basic language was developed and the types of applications that benefit from xHTML Basic programs. The basic programming structure of xHTML Basic is described along with the basic commands including links, images, and special symbols.

Introduction to SS7

ISBN: 1-9328130-2-0 Price: $14.99
Author: Lawrence Harte #Pages: Copyright Year: 2004

This introductory book explains the basic operation of the signaling system 7 (SS7). SS7 is the standard communication system that is used to control public telephone networks. This book will help the reader gain an understanding of SS7 technology, network equipment, and overall operation. It covers the reasons why SS7 exists and is necessary.

Creating RFPs for IP Telephony Communications Systems

ISBN: 1-9328131-1-X Price: $19.99
Author: Lawrence Harte #Pages: Copyright Year: 2004

This book explains the typical objectives and processes that are involved in the creation and response to request for proposals (RFPs) for IP Telephony systems and services. It covers the key objectives for the RFP process, whose involved in the creation and management of the RFP, and how vendors are invited, evaluated, and notified of the RFP vendor selection result. You will learn what are RFPs and RFQs and why and when companies use and RFPs for IP Telephony Systems.

ATM Basics

ISBN: 1-9328131-3-6 Price: $29.99
Author: Lawrence Harte #Pages: Copyright Year: 2004

Asynchronous Transfer Mode (ATM) is a high-speed packet switching network technology industry standard. ATM networks have been deployed because they offer the ability to transport voice, data, and video signals over a single system. The flexibility that ATM offers incorporates both circuit and packet switching techniques into one technology.

Althos Publishing, 404 Wake Chapel Road, Fuquay NC 27526 USA
1-919-557-2260 1-800-227-9681 Fax 1-919-557-2261 WWW.AlthosBooks.com

Worlds Largest Onlline Wireless Dictionary
WWW.WirelessDictionary.com

wireless Markup Language (WML)

ISBN: 0-9742787-5-0 **Price:** $34.99
Author: Bill Routt **#Pages:** 292 **Copyright Year:** 2004

Wireless Markup Language (WML) Scripting, Scripting and Programming using WML, cHTML, and xHTML explains the necessary programming that allows web pages and other Internet information to display and be controlled by mobile telephones and PDAs.

Introduction to Bluetooth

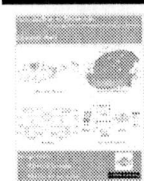

ISBN: 0-9746943-5-5 **Price:** $14.99
Author: Lawrence Harte **#Pages:** 60 **Copyright Year:** 2004

Introduction to Bluetooth explains what is Bluetooth technology and why it is important for so many types of consumer electronics devices. Since it was first officially standardized in 1999, the Bluetooth market has grown to more than 35 million devices per year. You will find out how Bluetooth devices can automatically locate nearby Bluetooth devices, authenticates them, discover their capabilities, and the process used to setup connections with them.

Introduction to CDMA

ISBN: 1-9328130-5-5 **Price:** $14.99
Author: Lawrence Harte **#Pages:** 52 **Copyright Year:** 2004

Introduction to CDMA book explains the basic technical components and operation of CDMA IS-95 and CDMA2000 systems and technologies. You will learn the physical radio channel structures of the CDMA systems along with the basic frame and slot structures.

Introduction to Mobile Telephone

ISBN: 0-9746943-2-0 **Price:** $10.99
Author: Lawrence Harte **#Pages:** 48 **Copyright Year:** 2004

Introduction to Mobile Telephone explains the different types of mobile telephone technologies and systems from 1st generation analog to 3rd generation digital broadband. It describes the basics of how they operate, the different types of wireless voice, data and information services, key commercial systems, and typical revenues/costs of these services. Mobile telephone technologies, systems, and services have dramatically changed over the past 2 years. tems to offer new services.

Introduction to Wireless Billing

ISBN: 0-9746943-8-X **Price:** $14.99
Author: Avi Ofrane, Lawrence Harte **#Pages:** 48 **Copyright Year:** 2004

Introduction to Wireless Billing explains billing system operation for wireless systems, how these billing systems are a bit different than traditional billing systems, and how these systems are changing to permit billing of non-traditional products and services. This book explains how companies bill for wireless voice, data, and information services.

Althos Publishing, 404 Wake Chapel Road, Fuquay NC 27526 USA
1-919-557-2260 1-800-227-9681 Fax 1-919-557-2261 WWW.AlthosBooks.com

Order Form

ORDER FORM

Phone: 919-557-2260
800-227-9681
Fax: 919-557-2261
404 Wake Chapel Rd., Fuquay-Varina, NC 27526 USA
Email: success@ALTHOS.com web: www.ALTHOS.com

Date: _____

Name: _____

Title: _____

Company: _____

Shipping Address: _____

City: _____ State: _____ Zip: _____

Billing Address: _____

City: _____ State: _____ Zip: _____

Telephone: _____ Fax: _____

Email: _____

Purchase Order # _____ (New accounts: please call for approval)

Payment (select): VISA ___ AMEX ___ MC ___ Check ___

Credit Card #: _____ Expiration Date: _____

Exact Name on Card: _____

Qty.	BOOK #	ISBN #	TITLE	PRICE EA	TOTAL

Book Total:	
Discounts:	
Sales Tax (North Carolina Residents please add 7% sales tax)	
Shipping: Please apply accurate shipping rates and surcharges per client and order.	
Total order:	

Worlds Largest Wireless Dictionary
WWW.WirelessDictionary.com

Providing Expert Information